Mathematical and Statistical Skills in the Biopharmaceutical Industry

Chapman & Hall/CRC Biostatistics Series

Shein-Chung Chow, Duke University School of Medicine
Byron Jones, Novartis Pharma AG
Jen-pei Liu, National Taiwan University
Karl E. Peace, Georgia Southern University
Bruce W. Turnbull, Cornell University

Recently Published Titles

For more information about this series, please visit: https://www.crcpress.com/go/biostats

Mathematical and Statistical Skills in the Biopharmaceutical Industry

A Pragmatic Approach

By
Arkadiy Pitman, Oleksandr Sverdlov
and L. Bruce Pearce

CRC Press
Taylor & Francis Group
Boca Raton London New York

CRC Press is an imprint of the
Taylor & Francis Group, an **Informa** business

A CHAPMAN & HALL BOOK

Contents

Preface

The current book is a product of three authors: the first author (AP) is a mathematician, the second author (OS) is a statistician, and the third author (LBP) is a pharmacologist. As a result, the book is slightly eclectic, since different parts represent views and experiences of different people.

The main subject of the book is application of mathematical and statistical skills in modern clinical drug development (mostly by biostatisticians). Most commonly, the development of new drugs is carried out by pharmaceutical or biotechnology companies. The whole process is a long, expensive, and complex enterprise, which should be performed keeping in mind such aspects as medical ethics, reproducibility and reliability of the results, efficiency in decision making (to maximize return-on-investment), and, of course, compliance with the regulatory standards.

Biostatisticians play a very important role in the drug development process—this is even documented in the regulatory guidelines. However, what does it really mean to be a biostatistician? Job descriptions may vary depending on the place of work, seniority level, etc. For instance, a biostatistican in a small one-drug company can be charged with numerous tasks, including data management, statistical programming (including data cleaning, analysis, and reporting), and biostatistics (trial design, submissions, publications, etc.) In a big pharma company, biostatistician may be focused more on biostatistics itself, while supporting a much larger number of studies and projects within the company portfolio. In contract research organizations (CROs), the work may be tailored to support the client's (big pharma) specific requests, which can be very broad. One can also mention academia and regulatory agencies, where biostatisticians play major roles as well, and their job descriptions are different from the ones in the biopharmaceutical sector.

Regardless of the place of work and assigned duties, a biostatistician typically has background in mathematics/statistics/biostatistics/computer science/data science/etc., and he or she has to solve various problems that arise in the context of drug development. Importantly, biostatisticians do not operate in a "vacuum"—they have to collaborate with many stakeholders including medical doctors, pharmacologists, clinical scientists, regulatory scientists, etc. As such, an important attribute for a biostatistician (in addition to technical background) is statistical consulting skills. In addition, the biopharmaceutical industry has a distinct feature—regulatory health requirements to ensure that research and development are carried out in full compliance with medical ethics, technical standards, and quality, to protect patients (present and future).

In this book, we describe a philosophy of problem solving based on a system of principles for pragmatic problem solving, specifically in the context of clinical drug development. The examples of applications vary from selection of necessary toolkits (Chapters 2 and 4) to providing help to a struggling neighbor (Chapter 3); from a general evaluation of the safety outcome of a randomized controlled trial (Chapter 5) to planning of open-ended projects (Chapter 7); from an attempt to salvage a failed clinical development program to some generalizations of encountered problems to the entire field and even entire drug development (Chapter 6). These ideas stem from the authors' many years of work in the biopharmaceutical industry. More specifically, the book will address the following aspects:

- A system of principles for pragmatic problem solving, which has been very helpful for the authors during their work.

- Some differences in the work of a biostatistician in small pharma and big pharma.

- Statistical programming and data management as the Atlas and Caryatid of biostatistics, especially in the small pharma setting.

- Some useful statistical background one may have just as an outcome of (graduate) studies that can be capitalized upon in the drug development enterprise.

- Some hot topics and current trends in biostatistics that are likely to be sustained in the future, explained in simple, non-technical terms.

- Applications of the described problem solving philosophy in a highly problematic transfusion field where many investigational compounds have failed.

- Application of the problem solving philosophy in planning of open-ended projects.

Clearly, no book can cover all important aspects. Some of the limitations (or rather intended omissions) of this book are:

- We do not cover pre-clinical/post-marketing activities. The presented problems are rooted in clinical development, and hence our focus is mainly on this part.

- We do not cover the work of biostatisticians in academia or regulatory agencies, since we have mostly theoretical knowledge about their work environment and work requirements.

- Our examples in this book are primarily from the transfusion field (based on joint experience of the first and the third authors).

• We do not discuss the challenges of career development in the corporate world, levels of competence, "soft" skills, lifecycles of enterprises/companies, etc., but focus mainly on mathematical/statistical applications and the pragmatic approach, as stated in the book title. The former issues have been discussed elsewhere. The authors' personal preferences are thought-oriented auditory authors like Laurence J. Peter (e.g. the Peter Principle) and Cyril Northcote Parkinson (e.g. Parkinson's Law).

We think the current book may be useful for biostatisticians working in the biopharmaceutical industry, as well as for graduate (Masters and PhD) students who major in biostatistics and/or related field and who consider a biostatistics-related career in this industry.

No special prerequisites are needed for reading this book—it is written in plain, non-technical language.

There are other books which provide a more comprehensive coverage of certain aspects of drug development. For instance, Stephen Senn's *Statistical Issues in Drug Development* [95] is an excellent time-tested guidance for pharmaceutical statisticians. Two other recent books, by Scott Evans and Naitee Ting, *Fundamental Concepts for New Clinical Trialists* [38], and by Arul Earnest, *Essentials of a Successful Biostatistical Collaboration* [32], also cover many important aspects related to biostatistician's role in clinical research. The book by Eric Holmgren, *Theory of Drug Development* [57], provides many useful insights into the nature of drug development from a mathematical perspective.

The roadmap for the current book is as follows:

Chapter 1 provides background for the pragmatic approach to problem solving. We identify vital parts of any problem solving process, and organize them in a system of principles that will be applied to tackle some real life problems from drug development. We also discuss some important differences between mathematics and statistics, and give some examples of proper and improper applications of statistics in evidence-based science, in particular in medicine and drug development.

Chapter 2 discusses statistical programming as a fundamental technical requirement for successful performance. We argue that while technological advances and increased specialization have led to separation of functions (even among statistical programmers, let alone biostatisticians), it is important to stay "hands on" and not to lose programming skills, especially when interfacing with complex trials/projects.

Chapter 3 discusses data management aspects, fundamental to the success of the clinical research. The provided examples highlight the complex, multi-disciplinary nature of the problems which require deep understanding of the neighboring fields and a high level of (intellectual) project coordination. For biostatisticians working in small pharma companies, this should provide some reassurance that they are not the only ones who have encountered this multi-hat, multi-problem environment where statistical programming, data management, and biostatistics constitute essentially the same thing. For people working in big pharma, this should provide an insight into what their collaborators (from the technical side) actually do at the "back end", and open the door for true collaboration.

Chapter 4 discusses biostatistical aspects from a perspective of someone working in big pharma. We highlight some challenges pertinent to the big pharma operating model, specifically due to a very large number of studies/projects that must be done in time, with a due level of technical rigor and accuracy, while maintaining compliance with the company standards and industry regulations. We elaborate on useful statistical skills and knowledge one may have based on their (graduate) coursework, try to elicit a "minimal sufficient" set of tools for the day-to-day work, and discuss some advanced topics in modern biostatistics which can be very useful in practice.

The common theme for Chapters 5, 6, and 7 is the application of the problem solving philosophy described in the earlier chapters.

Specifically, Chapter 5 describes a long-term effort of creating a new validated outcome scoring system for benefit/risk assessment. The authors would like to deliver a message that this is not "yet another scoring system" for assessing the outcome of an RCT, but rather a _thorough_ assessment of what the problem represents and how it could be solved, alongside with the actual solution.

Chapter 6 presents a case study of salvaging a failed clinical development program in the transfusion field. The attempt is to deliver several important messages, including the presence of non-covered gaps in modern regulations, the inability of standard tools to provide satisfactory solutions in non-standard situations, and the importance of careful planning of the entire development program to avoid disappointing failures in pivotal studies.

Chapter 7 is devoted to _realistic_ planning of open-ended projects. Due to the definition of an open-ended project as "a project that has never been done before and has no guarantee to be successfully finished", this problem becomes mostly of philosophical value, in the sense that the solution is very difficult to implement in practice, and it is almost impossible to sell. Nevertheless, dealing with such a problem allows demonstration of what can be done in such a desperate situation and an insight into whether it is possible to bring honesty in project planning at all.

Once again, the common ground for Chapters 5, 6, and 7 is the high-level strategic problem solving, which, as a rule, leads to various generalizations of relatively small tasks at hand, and helps better understand the world we are living in.

Acknowledgements:

We would like to thank the reviewers who provided their valuable comments/feedback at different stages of our work on this book: Vladimir Mats (Quartesian), Jens Praestgard (Novartis), Brian P. Smith (Novartis), Yevgen Ryeznik (Uppsala University), Stephen Senn (University of Sheffield), Lanju

Zhang (Abbvie), Francois Beckers (Merck Serono), Vladimir Anisimov (Amgen), as well as several anonymous reviewers appointed by the Editorial Board of CRC Press. We would like to sincerely thank the book's acquisition editor, John Kimmel, for his patience and support over more than three years of our work on this book.

I (the first author) would also like to thank everyone who worked either with or against me during my 20 years in drug development—it is not clear who contributed more to my world views. The special thanks go to my long-term boss (in all three incarnations of the HBOC company I worked for)—Mr. Zaf Zafirelis—who became a very close friend in the process of managing me. Without his trust, I would not be able to do what I have been doing since 2004; and without his tolerance of side projects that distract a valuable member of the management team for the past 3 years, this book would have never been written.

Authors

Arkadiy Pitman, M.Sc. has 20 years of broad based pharmaceutical and healthcare US experience with a strong background in mathematics, statistics, and logistics. Previously he had taught courses in mathematics, logic, computer sciences and statistics at various institutions in the Soviet Union, Ukraine and the US. He earned a Master of Science with Honors in mathematics at Kharkov State University, USSR in 1978. He currently works at HBO2 Therapeutics as Senior Director of Biostatistics and Data Management.

Oleksandr Sverldov, Ph.D. has worked as a statistical scientist in the biopharmaceutical industry since 2007. He is currently neuroscience disease area statistical lead in early clinical development at Novartis. He has been involved in active research on adaptive designs for clinical trials to improve efficiency of drug development. He edited a book *Modern Adaptive Randomized Clinical Trials: Statistical and Practical Aspects* which was published by CRC Press in 2015. His most recent work involves design and analysis of proof-of-endpoint clinical trials evaluating digital technologies.

L. Bruce Pearce, Ph.D. has worked as a pharmacologist and toxicologist in the area of small molecule and biologics drug development for more than 25 years. His career began as an academic teaching graduate and medical students. His introduction to drug development was as an entrepreneur, and he subsequently moved to the development of blood substitutes where he first collaborated with Arkadiy Pitman. Since 2009 he has served as a consultant to very early and late stage biotechnology and pharmaceutical companies for the development of small molecule-based drugs, biotechnology-derived natural and recombinant biotherapeutics, and medical devices. Areas of interest have included, pharmacology, toxicology, pharmacokinetics, risk assessment, biocompatibility, due diligence, early and late stage regulatory strategy, preIND, IND, 510k, 505(b)(1) and 505(b)(2) NDAs and BLAs covering a broad spectrum of therapeutic areas.

List of Abbreviations

ADAM: Analysis Data Model

AE: Adverse Event

AUC: Area Under the Curve

BLA: Biological License Application

BLA: Benefit Risk Assessment

CBER: Center for Biologics Evaluation and Research

CDISC: Clinical Data Interchange Standard Consortium

CDER: Center for Drug Evaluation and Research

CDRH: Center for Devices and Radiological Health

CHMP: Committee for Medicinal Products for Human Use

CRF: Case Report Form

CRO: Contract Research Organization

DM: Data Management

DMC: Data Monitoring Committee

ECD: Electronically Captured Data

EMA: European Medicines Agency

FDA: Food and Drug Administration

Hb: Hemoglobin

HBOC: Hemoglobin-Based Oxygen Carrier

HCT: Hematocrit

ICH: International Conference on Harmonisation

IND: Investigational New Drug

ISE: Integrated Summary of Efficacy

ISS: Integrated Summary of Safety

MedDRA: Medical Dictionary for Regulatory Activities

MTD: Maximum Tolerated Dose

NDA: New Drug Application

OSS: Outcome Scoring System

PK: Pharmacokinetics

PD: Pharmacodynamics

PoC: Proof-of-Concept

pRBC: packed Red Blood Cells

R&D: Research and Development

RAR: Response-Adaptive Randomization

RBC: Red Blood Cells

RCT: Randomized Controlled Trial

RSS: Royal Statistical Society

RWE: Real World Evidence

SAE: Serious Adverse Event

SAP: Statistical Analysis Plan

SDTM: Study Data Tabulation Model

SOC: System Organ Class

SOP: Standard Operating Procedure

SP: Statistical Programming

THb: Total Hemoglobin

VA: Veterans Affairs

WHO: World Health Organization

1

Background and Motivation

As the title suggests, this book will discuss a pragmatic approach to problem solving in modern clinical drug development. We shall emphasize some basic yet powerful principles that can help leverage the biostatistician's value.

We start this chapter with a general discussion of the philosophy of problem solving (§1.1–1.2), followed by a discussion of mathematics and statistics as important tools for problem solving (§1.3). Furthermore, we give a brief discussion of the modern drug development process (§1.4) and revisit some issues in the application of statistics in evidence-based science (§1.5). A summary of what this book is all about is given in §1.6.

1.1   Pragmatic approach to problem solving

"First, you select the problem you would like to solve, then you list at least ten reasons why this has not been solved. But in picking that problem be sure to analyze it carefully to see that it is worth the effort. It takes just as much effort to solve a useless problem as a useful one. Make sure the game is worth the candle."

Charles F. Kettering "Research is a State of Mind"

One may think that any problem solving is pragmatic by definition—after all, it is the hunt for a practical solution. If we include into the definition of a "practical solution" the decision making on whether you spend time on the problem along with reasoning for such a decision, the words "pragmatic approach" become completely unnecessary. It is interesting that the actual short-term practicality might be unrelated to real pragmatism. For example, the short-term practicality of modern mathematics is highly questionable, but the long-term pragmatism in development of fundamental science is undisputable. One of the greatest philosophers, Stanisław Lem (who happened to be a famous science fiction writer as well), in his 1964 philosophical tractate *Summa Technologiae* coined the term "crazy tailor" while describing modern pure mathematics. According to Lem, mathematics is a tailor shop where

1

clothes for any imaginable creatures are made in advance: when these creatures arrive, the clothes are already available.

Now, seriously: the gap between creation of tools and their first applications for pure mathematics has been increasing since the split of pure and applied mathematics, which is now estimated as 2–3 centuries. Simply put, current applications are based on theories developed before 1800, and it is expected that cutting edge theories which pure mathematicians are working on now will be needed only after 2200. Worse still, about 95% of the studied structures might not be used at all. We think that is the main reason why pure mathematicians are not considered to be "truly pragmatic" and most statisticians want to separate themselves from pure mathematicians.

There is an interesting psychological phenomenon: mathematicians are considered to be the best problem solvers but are deemed unpractical. If we rephrase the title from the epigraph, we can say that *"Problem Solving is a State of Mind"*, and mathematicians, mostly due to their training, are recognized as having it, thus making them superior in this aspect. Perhaps in order to balance this aspect, and to have an opportunity to deny their findings, others have developed a superiority complex on practicality. In fact, pure mathematicians are probably the most practical/pragmatic people in the world, since they constantly have to decide whether "the game is worth the candle" while working on a complex problem (take Grigori Perelman[1] as an example). The only "impractical" quality of mathematicians (as well as any true problem solvers) is the pursuit of a problem just for moral satisfaction. Strictly speaking, though, the decision to continue work on a problem "just for fun" could be defended as pragmatical, especially if it is done during one's leisure time. After all, it is great entertainment, plus you never know whether newly acquired knowledge and/or skills could be useful elsewhere.

1.2   Problem solving skills

"Expertise in one field does not carry over into other fields. But experts often think so. The narrower their field of knowledge the more likely they are to think so."

Robert A. Heinlein "The Notebooks of Lazarus Long"

The quoted character (Lazarus Long) is perhaps one of the most pragmatic persons described in literature. He was created to demonstrate the idea that a combination of common sense, a good brain, proper education and pragmatism

---

[1] The one who proved the Poincaré conjecture (one of the seven Millenium Prize Problems stated by the Clay Mathematics Institute on May 24, 2000).

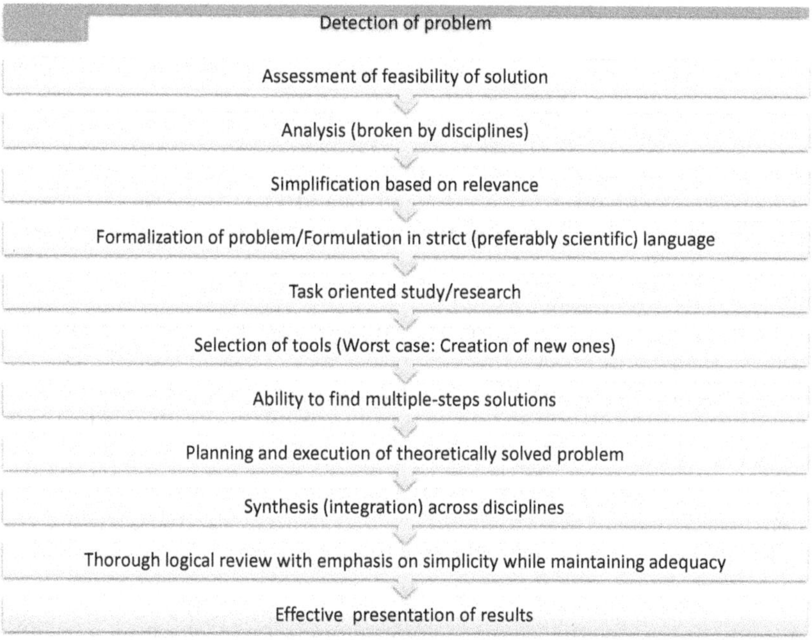

Detection of problem
Assessment of feasibility of solution
Analysis (broken by disciplines)
Simplification based on relevance
Formalization of problem/Formulation in strict (preferably scientific) language
Task oriented study/research
Selection of tools (Worst case: Creation of new ones)
Ability to find multiple-steps solutions
Planning and execution of theoretically solved problem
Synthesis (integration) across disciplines
Thorough logical review with emphasis on simplicity while maintaining adequacy
Effective presentation of results

FIGURE 1.1
Vital parts of a problem solving process

is the best recipe for successful problem solving in real life and actually could even serve as a decent substitute for morality and religion. The meaning of the epigraph is pretty clear: highly specialized professionals who limit themselves to narrow fields are useless for anything else.

One important direct consequence of the above quote can be formulated as follows: *"Expertise is not transferable, but skills and approach are"*. Indeed, real-life problem solving skills and a creative approach can be developed anywhere (in any field)—the bigger the number and diversity of these fields, the stronger the skills are, and the easier application in a new situation is.

The first author's firm philosophy is that relevance of past experience depends much more on what you have carried over into today's work rather than on where you have worked.

Let us take a closer look at what is meant by "problem solving". Figure 1.1 displays elements (may not be all inclusive) which represent vital parts of any problem solving process. Most of the elements are self-explanatory and we shall just add a few clarifying notes. *Detection* is probably the hardest part. A lot of problems are not solved simply because they are not detected. *Feasibility* is two-fold, and it is in fact a part of detection: some problems are indeed non-solvable, while others are left unattended because they are considered non-solvable. *Analysis* (broken down by disciplines) implies that all problems are multidimensional; and yet a successful analysis does not necessarily guarantee a successful synthesis. *Simplification* (or principle of parsimony) is another critical piece. We have a good company here: Einstein's take on Occham's Razor—*"As simple as possible, but not simpler."* Going "outside the box" and over-simplification are again two sides of the same coin. *Formalization* of the task is an art by itself; *selection of tools* is certainly essential and, in fact, new tools should generally be considered as a last resort (unavoidable evil). *Planning and execution* of the discovered solution can be a problem in itself; and finally, the *effective presentation of results* and ability to sell both expertise and solution are very crucial.

By and large, the toolkit displayed in Figure 1.1 is used by any person who has been engaged in problem solving either at work or in everyday situations. While the goal of the current book is to demonstrate how the approach works in clinical drug development, some ideas can be successfully applied in other settings as well.

Let us now look at the requirements for a professional problem solver— a person who does it for living and who is called upon only when a serious problem is detected and all attempts to solve it from the inside have failed. The success drivers are the *logic* and the *logistics*. The important prerequisites include understanding the elements and structure of logical thinking, the ability to learn "from scratch" quickly and deeply in an unfamiliar field, and the ability to focus on what is most relevant (accurate assessment of scope, applicability, limitations, precision and practicality). After all, goal-oriented practicality, simplicity, comprehensiveness and acceptability frequently take precedence over the beauty of science.

In any case, either for an amateur or a professional, it is good to remember that problem solving is more than work; it is rather a passion or even a lifestyle. The major difference between casual and professional problem-solving is grounded in the organizational aspects that are necessary for a professional. Theoretically it is possible to be a genius problem solver without any political and organizational skills; however, in practice, such individuals usually have a very hard time in self-realization and, as a rule, remain unknown and unaccepted.

One of the major goals of this book is to discuss not only solutions to the problems, but also attempts for getting acceptance as well. Considering that drug development is far away from pure science, achieving acceptance is, in many cases, one of the most complicated parts of the solution.

1.3   Mathematics versus statistics

"A mathematician believes that $2 - 1 \neq 0$; a statistician is uncertain."

Anonymous

Broadly speaking, science can be classified into three major categories: natural sciences (the study of natural phenomena), social sciences (the study of human behavior and society), and formal sciences (mathematics). The first two types are *empirical* sciences in which new knowledge is acquired through observation of certain phenomena. By contrast, in mathematics new results are obtained by proving certain conjectures (theorems) using formal logic, a system of axioms, and previously established results. In the context of empirical sciences, one can distinguish three types of knowledge, in the order of increasing complexity: 1) descriptions of phenomena in terms of observed characteristics of the events; 2) description of association among phenomena; and 3) descriptions of causal relationships among phenomena [88]. Many of scientific activities aim at establishing causal relationships between various phenomena, ideally using mathematical equations.

In natural sciences such as physics or chemistry, mathematics has been a very powerful tool to describe causal relationships and, therefore, enable prediction. The physicist Eugene Wigner (1902–1995) in his 1960 seminal paper [103] wrote:

> "The miracle of the appropriateness of the language of mathematics for the formulation of the laws of physics is a wonderful gift which we neither understand nor deserve. We should be grateful for it and hope that it will remain valid in future research and that it will extend, for better or for worse, to our pleasure, even though perhaps also to our bafflement, to wide branches of learning."

While many physical laws have been successfully described using mathematical models, such models are only approximations (albeit quite accurate) to reality. This is captured in Albert Einstein's (1879–1955) famous quote:

> "As far as the laws of mathematics refer to reality, they are not certain; as far as they are certain, they do not refer to reality."

In biological and social sciences the formulation of cause–effect relationships is much less straightforward than in natural sciences. There is a substantial random variation among humans with respect to their physiology, environmental and lifestyle factors which makes it difficult, if not impossible, to draw definitive conclusions about observed phenomena. For instance, in epidemiology the goal is to determine how certain risk factors affect the incidence of a disease in a population. Does smoking cause lung cancer? If so, then why are there smokers who do not develop lung cancer in their lifetime? Clearly, there is a strong association between smoking and lung cancer, but the causal link is elusive. As soon as we get away from deterministic relationships, establishing "true" causality is practically impossible.

In 1965, the English statistician Sir Austin Bradford Hill (1897–1991) (who is also known as the designer of the first randomized clinical trial evaluating the use of streptomycin in treating tuberculosis in 1946) formulated nine criteria that can be used to provide *epidemiological* evidence of a causal relationship between risk factors and disease [56]. These criteria include: 1) strength of association (effect size), 2) consistency of the findings (reproducibility), 3) specificity, 4) temporality (the effect must be observed after the cause, possibly with delay), 5) biological gradient (greater exposure should magnify, or sometimes lessen, the incidence of the disease), 6) plausibility, 7) coherence, 8) experiment ("occasionally it is possible to appeal to experimental evidence"), and 9) analogy. While Bradford Hill's criteria have been widely used in epidemiology, they are still subject to scientific debate.

The challenges of building mathematical models to describe phenomena in economics, social sciences, and even technology were also acknowledged by the originator of cybernetics Norbert Wiener (1894–1964). In his work "God & Golem, Inc." [102], the famous researcher wrote:

> "The success of mathematical physics led the social scientist to be jealous of its power without quite understanding the intellectual attitudes that had contributed to this power. The use of mathematical formulae had accompanied the development of the natural sciences and become the mode in the social sciences. Just as primitive peoples adopt the Western modes of denationalized clothing and of parliamentarism out of a vague feeling that these magic rites and vestments will at once put them abreast of modern culture and technique, so the economists have developed the habit of dressing up their rather imprecise ideas in the language of the infinitesimal calculus... Difficult as it is to collect good physical data, it is far more difficult to collect long runs of economic or social data so that the whole of the run shall have a uniform significance."

In the authors' opinion, what Wiener had in mind is simple advice to not expect results of deterministic quality (similar to gas laws in thermodynamics) while applying statistics anywhere outside of the areas of theoretical physics operating with a *truly astronomical* number of identical elements. Even the authors' beloved science fiction writer Isaac Asimov (1920–1992) had to create the Second Foundation to correct the truly deterministic results of Psychohistory that was devised to predict the future in a sociological and historical sense (and/or science).

Jerzey Neyman (1894–1981) in his 1957 paper on "inductive behavior" as a basic concept of philosophy of science [75] describes some limitations of deterministic models and motivates the statistical (frequentist) approach to many scientific fields. The main difference between mathematical (deterministic) and statistical (probabilistic) reasoning is that mathematics uses deduction and formal logic in pursuit of the definitive answer to a problem, whereas in statistics, inferences are made based on observed data using inductive arguments, and the conclusions are never certain because data are random.

Unlike mathematics which is grounded on formal logic, statistics does not have a single philosophy that forms the basis for scientific reasoning. At least three different statistical paradigms can be distinguished: 1) Likelihood-based (Fisherian) paradigm; 2) Classical/error statistics (Neyman–Pearson) paradigm; and 3) Bayesian paradigm. Applications of these paradigms are context-specific.

The likelihood-based paradigm (due to R. A. Fisher (1890–1962)) has been widely used as a tool to address evidence questions, in particular through hypothesis testing. A researcher formulates a null hypothesis (H_0) to be tested using an outcome of interest. Once experimental data have been collected, the researcher computes *P-value*, the probability of observing a result at least as extreme assuming that H_0 is true. The P-value provides a degree of confidence in H_0 in light of the observed data—small P-value indicates that data are inconsistent with the assumption that H_0 is true. Never can H_0 be proved ("Absence of proof is not proof of absence"). R. A. Fisher himself acknowledged that such a hypothesis testing approach should be used only if very little is known about the problem at hand, and that the P-value should be only communicated, but not used to make a Yes/No decision.

The Neyman–Pearson approach (due to Jerzey Neyman (1894–1981) and Karl Pearson (1857–1936)) was developed to introduce scientific rigor into the decision making process. This approach is well suited in *quality control* settings where experiments are repeated multiple times and where it makes sense to control the "long-run" false positive and false negative probabilities below certain thresholds. In this case, a decision (acceptance or rejection of one of the two competing hypotheses) is the primary goal. Importantly, acceptance/rejection of a hypothesis has an implication to take a certain action (decision) and not to claim that a hypothesis is "true" or "false".

There are some fundamental differences between the Neyman–Pearson and the Fisherian approaches to hypothesis testing, which even led to tension between the researchers. In many social, biological and natural sciences (in particular, in clinical trials) a hybrid of two theories has been widely used, which has also been widely criticized [49, 50]. One thing that Neyman, Pearson and Fisher all agreed upon is that statistics should never be used mechanistically. Some recent discussions further highlight the problems with the use of statistical significance as the basis for making scientific inferences [2, 51, 99, 100].

The Bayesian paradigm (named after Thomas Bayes (1701–1761)) is another powerful approach to statistical inference. With this approach, the probability is viewed as a state of belief or current state of knowledge (in contrast to the frequentist interpretation of probability as a limit of relative frequency of event occurrence in a large number of independent trials). A prior probability is combined with the likelihood (data) using Bayes' formula to give a posterior probability. One advantage of the Bayesian approach is that it utilizes historical data (via prior distributions) and allows direct probabilistic statements about parameters of interest. A disadvantage is the subjectivity component (different priors can lead to different results). However, there are many problems where the Bayesian approach fits well whereas frequentist methods cannot provide an adequate solution [1].

One of the great statisticians of the 20th century George E. P. Box (1919–2013) once said: *"Statisticians, like artists, have the bad habit of falling in love with their models."* Given different approaches to statistical inference and possibly different solutions to the questions of interest, what should be the criteria for statistics to qualify as science? Some of our (subjective) ideas in this regard are as follows:

- When we carefully plan an experiment (utilizing knowledge from relevant scientific fields) and define the scope of data to be collected (necessary to address the research questions of interest).

- When, while performing the experiment, we acquire high quality data that are sufficient to support statistical modeling at the analysis stage.

- When we give a clear description of how the experiment was performed, what the assumptions were, and what the limitations are.

- When we use pre-specified, not data-driven, decision rules.

- When we look at the body of evidence, not just the results of a single experiment.

1.4   A look at modern drug development

"...The point is that the successful, useful drugs outnumber losers. For every loser—Thalidomide, Selacryn, Montayne, Oraflex, Bendectin; those and the other few failures you hear about on TV news and '60 Minutes'—there have been a hundred winners. And it isn't just the pharmaceutical companies who are gainers. The big winners are people—those who have health instead of sickness, those who live instead of die..."

Arthur Hailey "Strong Medicine"

The quote in the epigraph is taken from a 1984 novel by Arthur Hailey, where the author masterfully described challenges of pharmaceutical drug

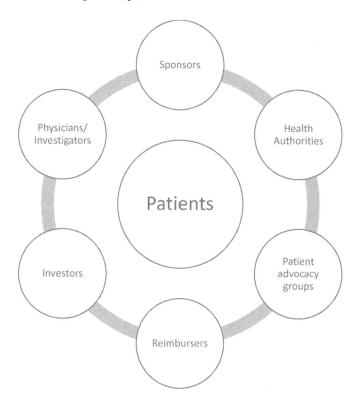

FIGURE 1.2
Key stakeholders in the clinical drug development process

development and interactions with the FDA in the period from late 1950's to early 1980's. The gentleman (who is a practicing medical doctor) assures his wife (who is a big pharma executive) that the pharmaceutical industry (despite all its faults and critiques) does bring big value to the mankind. The reference to the "losers" and the "winners" is made presumably in the context of the *already approved* drugs. The "losers" are ones that have been initially approved, yet later exhibited terrible side effects and had to be withdrawn from the market. It is noteworthy that there were cases when some of the withdrawals were reversed after backlash from clinicians (for possibly different indications).

Now, 35 years since this novel was published, the pharmaceutical drug development and the regulatory landscape has evolved tremendously. In contemporary society, advances in science, technology and medicine have led to significant improvements in standards of healthcare and quality of human life. However, many diseases still present significant unmet medical need. The research and development of therapies for treatment and prevention of complex diseases is a long, daunting, expensive, and risky process [20]. Before describing the drug development process in detail, let us outline its key stakeholders (Figure 1.2).

Patients are the key stakeholders, because the fundamental objective of drug development is to deliver safe and efficacious drugs to those with the medical need, thereby extending and improving human life. Here it is important to distinguish between *clinical practice* and *clinical research*. The goal of clinical practice is to benefit a specific patient, whereas the goal of clinical research is to advance medical knowledge to benefit future patients. Clinical research is carried out through randomized controlled trials (RCTs), which have been recognized as the gold standard for clinical investigation. Importantly, both the clinical trial participants and the future patients should expect to benefit from the RCT investigation. The principle of human research ethics (e.g. the Declaration of Helsinki) states that "...*the subject's welfare must always take precedence over the interests of science and society, and ethical considerations must always take precedence over laws and regulations.*" At the same time, careful scientific experimentation is required to draw conclusions beyond the clinical trial and generalize the results to a population of patients with the disease. These two competing requirements are referred to as *individual* and *collective* ethics, and interplay between them is never simple [81].

Sponsors are research organizations (e.g. pharmaceutical/biotechnology companies or non-profit organizations) that undertake clinical research. Drug development is a very expensive process which requires massive investment of capital into both human resources (qualified scientists, research personnel, etc.) and materials (research laboratories, equipment, etc.) From the sponsor's perspective, the R&D investment should be justifiable and should generate revenue in the future. Sponsors can also borrow necessary expertise by collaborating with academic institutions such as research universities.

Health Authorities (regulatory agencies) are tasked with ensuring that investigational drugs for clinical trials and for commercialization are safe, efficacious and high quality. Different countries have different sets of regulatory requirements; some of these requirements are legally enforced and others are non-legally binding guidelines. For instance, in the United States the legal framework is the US Code of Federal Regulations (CFR) which is carried out by the Food and Drug Administration (FDA). The FDA has the Center for Drug Evaluation and Research (CDER) which is responsible for review and approval of Investigational New Drug (IND) applications and New Drug Application (NDA) submissions.

Physicians will eventually prescribe new medicines to their patients. Physicians should constantly stay on top of most recent advances in clinical research, to ensure best available standard of care for the patients. Sponsors frequently contract with research sites and private medical practices to condust clinical trials. In this case, practicing physicians act as *investigators*.

Reimbursers are the organizations that will grant reimbursement once the drug is marketed. Some countries such as Australia and Canada have an ex-

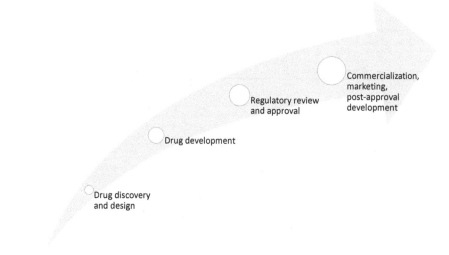

Commercialization,
marketing,
post-approval
development

Regulatory review
and approval

Drug development

Drug discovery
and design

FIGURE 1.3
Major stages in development of a new drug

plicit requirement of evaluation of cost-effectiveness of a new treatment before granting reimbursement [76].

Investors are also important stakeholders as their decisions to invest into pharmaceutical/biotech companies have a direct impact on the clinical research enterprise.

Patient advocacy groups have an increasing stake in drug development. It is argued that clinical trialists should do a better job communicating with these and other external parties while the trials are ongoing and the definitive results have not been obtained [27].

1.4.1 Stages of drug development

Most commonly the development of new drugs is carried out by pharmaceutical or biotechnology companies. Within a company, drug development teams are formed to work on particular projects within a particular therapeutic area. Each drug development team consists of subject matter experts from different scientific disciplines. Within a team there may be sub-teams representing different development streams such as chemical and pharmaceutical development, nonclinical development, clinical development, etc. A clinical development sub-team typically includes representatives from trans-

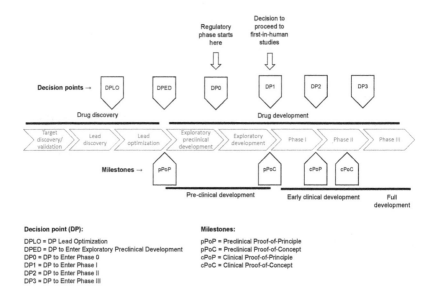

FIGURE 1.4

Drug discovery and drug development processes

lational research, clinical medicine, clinical pharmacology, clinical operations, biostatistics, regulatory affairs, pharmacovigilance, finance, etc. A close collaboration and efficient communication among various functions is key. The scientific and operational efforts are coordinated by a global project team leader.

Figure 1.3 shows major stages in development of a new drug. The first two stages, *drug discovery and design* and *drug development* are broken down into more steps, displayed in Figure 1.4. Different companies may utilize different variations of these strategies. Importantly, the drug discovery process is characterized by high degrees of freedom, whereas the drug development is strictly regulated.

The drug discovery starts with the *target discovery*—identification of the right biological target(s) that is(are) linked to the disease of interest. In the *lead discovery* phase, multiple candidate compounds that modulate an activity of the biological target are screened to identify the lead series. In the *lead optimization* phase, the goal is to identify one or more lead compounds to progress into drug development. Pre-clinical evidence of target validation is essential to facilitate the decision to proceed with the development of a compound.

Once the lead compound has been identified, it is progressed through various stages of development, which can be broadly categorized as *pre-clinical* development, *early clinical* development (phase I and phase IIa), and *full clinical* development (phase IIb and phase III). In the pre-clinical development, the objectives are: 1) to manufacture a drug product that is high quality and has acceptable bioavailability of the active ingredient to exert its therapeutic effect; and 2) to study drug efficacy, safety, and pharmacokinetics in animals and in laboratory settings.

If the drug is found to be acceptably safe and have acceptable pharmacokinetic properties and beneficial therapeutic effect in pre-clinical studies, a decision to enter the clinical development phase (DP1) is made. At this point, the sponsor would submit an Investigational New Drug (IND) application to the FDA to notify the agency on the intention to pursue the clinical development of the drug.

The clinical stage of drug development starts with phase I studies in humans to explore safety, tolerability and pharmacokinetics (PK) of the compound and to identify doses that are suitable for testing in subsequent studies. Phase I trials are typically conducted in healthy volunteers; however in life-threatening diseases such as cancer, the drugs are highly toxic and for ethical reasons such trials are conducted in patients who have failed standard treatment options. In addition to safety, tolerability and PK, phase I trials assess early signals of efficacy using relevant biomarkers. At the end of phase I, if the results in healthy volunteers are encouraging, a decision to enter phase II of development (DP2) is made.

Phase II can be categorized into two parts: phase IIa (early clinical development) and phase IIb (full clinical development). In phase IIa the goal is to demonstrate the clinical proof-of-concept (PoC) by showing that the candidate drug exerts promising therapeutic effect in the target patient population. In phase IIb the goal is to characterize the therapeutic dose range by estimating the dose–response profile and identifying a dose (or doses) with most promising benefit/risk properties for testing in subsequent confirmatory trials. Upon successful completion of phase II, the decision to enter phase III of development (DP3) is made. This is a major milestone for the sponsor, and it is associated with a substantial investment of capital.

Phase III clinical trials are confirmatory studies in a large and diverse group of patients with the disease of interest. These trials are specifically designed to test clinical research hypotheses. Most of phase III trials are randomized, controlled, double-blind studies testing superiority of the drug over placebo or a standard of care. Based on phase III trial results, hopefully a conclusion can be made that the experimental drug is safe and efficacious in treating the disease and can improve the quality of life. In this case, a decision can follow to prepare a *regulatory dossier* (documents summarizing all available pre-clinical and clinical data from conducted studies) to submit to health authorities in request of marketing authorization. In the US, a regulatory dossier is referred to as a New Drug Application (NDA), which is submitted to the FDA[2].

Even if the drug receives regulatory approval and is marketed, there are further clinical studies (phase IV trials) to assess long-term safety in large populations of patients.

Any drug development enterprise is very expensive and risky. Unfortunately, in the past two decades increases in R&D expenditures have not led to increased productivity [94]. A recent analysis of R&D productivity showed that *"the industry is not yet 'out of the woods' and ...many of the systemic issues affecting pharmaceutical R&D productivity are still being resolved"* [70]. Some strategies to improve efficiency of drug development have been proposed [13, 66]. Health authorities have also been acknowledging the challenges and have launched various initiatives to modernize drug development [43, 44].

1.4.2 Factors that have had an impact on drug development

The drug development enterprise has been evolving over years [25, 26]. Here we briefly describe some factors (a subjective selection) that have emerged and triggered innovation and changes. These include: 1) development of personalized medicine; 2) development of digital technologies; 3) globalization; 4) changes in regulatory landscape; and 5) the biostatistics discipline.

Recent developments in biotechnology and genomics have led to better understanding of the biological basis of many fatal diseases such as cancer, which, in turn, has led to paradigm shifts in clinical trial design and analysis, with an increasing focus on development of *personalized medicine* [96]. A traditional clinical trial tests a hypothesis of an average treatment effect in a large population. Personalized medicine trials test hypotheses in well-defined subgroups of patients who are likely to benefit from the drug. Don Berry has a nice paper describing the "Brave New World" of clinical oncology which can fundamentally change the drug development paradigm [11]. He argues: *"...No one understands the regulatory model in the Brave New World. And no one understands the corresponding business model for pharmaceutical companies. All that is clear is that both will be different from today's."*

The development of digital technologies (Internet, social networks, web-based data capture systems, and smartphones) has enabled capturing, storing and handling massive amounts of clinically relevant data [72, 87]. The implications for change in clinical trial practice are clear—potential for enhanced data quality, opportunities for automation of many administrative processes, gains in study efficiency and accuracy, etc.

---

[2]While discussing the *drug* development and the FDA's CDER as the division that reviews NDAs, it should be mentioned that there are other divisions of the FDA tasked with reviewing clinical trial results and applications for marketing authorization; e.g. the Center for Biologics Evaluation and Research (CBER) reviews Biological License Applications (BLAs); the Center for Devices and Radiological Health (CDRH) reviews medical device applications, etc.

As the world is becoming more global, so is drug development. Clinical trials become increasingly multi-center and multi-regional. This calls for harmonization of standards that are accepted internationally. The International Conference on Harmonization (ICH), organized in Brussels in 1990, brings together representatives from health authorities and the pharmaceutical industry in Europe, Japan and the US, to harmonize technical requirements for the safety, efficacy and quality of drugs and to allow efficient drug registration.

There have been continuous changes in regulatory requirements for the new medicines. Health authorities have been constantly raising the bar for data quality and scientific rigor of drug development programs. Numerous new regulatory guidelines have been issued to reflect these initiatives. Incentives for developing drugs for certain high unmet medical needs (such as rare diseases) have attracted companies to invest more into this area.

Finally, there is an increasing recognition that *biostatistics* (in the broad sense), can significantly contribute to the success of drug development which is certainly very exciting for the profession.

1.5 Statistics and evidence-based science

"If you torture the data enough, nature will always confess."

Ronald Coase [24]

The purpose of this section is to revisit some applications of statistics as a research tool, in particular in the context of clinical trials. The ideas presented here are, of course, subjective and can be rightfully challenged on numerous grounds.

In recent years we have seen a triumphant walk of statistics. It is used everywhere—millions of studies are conducted, some things that were believed to be true are confirmed, some common myths are debunked. All this happens due to a tool that has been a legitimate basis for scientific development: evidence-based science.

It should be noted that in early years statistics was used mostly as a descriptive tool. Natural sciences (physics, chemistry, etc.) generally considered statistics as an auxiliary instrument, but not as an engine to acquire new knowledge. Unfortunately, the use of statistics as an ultimate proof of underlying theories is not uncommon, in particular in medicine and drug development. It stands to reason that statistics alone cannot substitute the real scientific research, which is based on many things, including thinking, logic, application of already acquired knowledge and which is definitely not limited to the experiment and black-box modeling. One example where mathematical/statistical modeling is widely used as a tool for making predictions is the finance industry. The crash of financial derivatives in the late 2000's indicate that models based on past behavior are not necessarily good for predicting the future.

To substantiate our position, we use the following quote from a 2018 paper by a group of researchers who responded to some recent recommendations on redefining the statistical significance threshold [68]:

> "...Research should be guided by principles of rigorous science, not by heuristics and arbitrary blanket thresholds. These principles include not only sound statistical analyses, but also experimental redundancy (for example, replication, validation, and generalization), avoidance of logical traps, intellectual honesty, research workflow transparency and accounting for potential sources of error. Single studies, regardless of their P-value, are never enough to conclude that there is strong evidence for a substantive claim. We need to train researchers to assess cumulative evidence and work towards an unbiased scientific literature. We call for a broader mandate beyond P-value thresholds whereby all justifications of key choices in research design and statistical practice are transparently evaluated, fully accessible and pre-registered whenever feasible."

The following examples demonstrate how things can go wrong with improper use of statistics in clinical drug development.

#1. Confirmation or rejection of underlying theory

The major advantage of a prospectively designed randomized controlled trial (RCT) is thought to be a confirmation or rejection of an underlying theory. Not so easy. Suppose there are several competing theories that predict the same outcome, but the RCT designer is aware of only one. For him the results of his experiment are an ultimate proof of the theory in question. Somehow any alternative interpretation is doomed as inferior just because the first theory was prospectively formulated first. If this looks familiar to you, you are right—this has happened many times. One of the most prominent examples is the "confirmed by experience" hypothesis that the Sun revolves around the Earth. Any alternative explanation was doomed as heresy, since "eyes don't lie". Notice that in this case being first is the most important part of the prospectively designed experiment.

#2. Misconception that the RCT results are easily understood

There is a belief that the RCT results (because they can be formulated in simple numerical terms) are easily understood by anyone. This belief is widely used by self-appointed patient advocates who request to make all data, including intermediate results, available to the public. Moreover, some well-executed campaigns were based solely on personal preferences/opinion and were supported by highly problematic application of the most elaborate statistical tools (e.g. meta-analysis). An interested reader is referred to the *Manual for Successful Crusade in Defense of Patients' Rights* (cf. Appendix B) for a recipe for such crusades in defense of patients. This was published by the first author in a Russian medical journal (under the section "Scientists Are Joking"), jointly with one of the leading Russian hematologists (Head Hematologist of the Russian National Research Medical University named after N. I. Pirogov in Moscow). The original article is English-translated for convenience.

Jokes aside, such campaigns can be very harmful. For successful, already approved drugs it could be costly to demonstrate that the drug is not as bad as some "have proven" (usually after recall, tense communications with regulators; sometimes an extra confirmatory study and then getting back to the market with a blemished reputation). For not yet approved drugs (that are still under development), the results can be devastating—the entire field can be demolished without a reason, and the development of promising, potentially life-saving therapies can be terminated. The first and the third authors happened to be at the receiving edge of such a campaign in both described scenarios. The drug Hemopure was already approved in South Africa (SA) and not approved in the US, where the hemoglobin-based oxygen carrier (HBOC) field was essentially destroyed (despite loud protests from practicing physicians) as a result. (The journey described later in Chapter 6 became much longer and much harder than it would have been without such a campaign; thus this entire passage could be viewed by an objective reader as a *heavily biased opinion*.)

Let us get back to the Earth and the Sun relationship. If you asked 1 million randomly selected people what was the essence of conflict between Galileo Galilei and the Church, you would probably get an overwhelming majority (out of the people who know who Galileo was) stating that it was whether the Sun revolves around the Earth or the Earth revolves around the Sun. Some would even provide the last words of the great scientist *"But it (the Earth) still moves."* In fact, the scientific essence was whether the Sun revolves around the Earth or the Earth revolves around itself. [3]

#3. Misconceptions exist even on the professional level

Most of professionals understand the difference between the model-fitted equation and the truth. If the data from legendary (non-confirmed by history) Galileo experiments in Pisa were given to a modern statistician, it would probably yield a linear or an exponential fitted formula, but for sure not $S = gt^2/2$. On the other hand, there is a belief that qualitative questions could be answered after statistical evaluation of the experimental results. If

---

[3] It was thought that the Sun makes a circle in 24 hours, which "was seen" during the day as the Sun moved through the sky. The alternative solution (proposed by Galileo Galilei) suggested that the Earth revolved around itself in 24 hours. The revolution of the Earth around the Sun (in 365 days) has nothing to do with this discussion.

modern mathematics was applied to actual results, Aristotle instead of Galileo would have been proven right: heavy objects do fall faster (due to air resistance that nobody could describe properly at that time). Generally, we are back to example #1 with an alternative explanation based on hidden covariates.

These examples illustrate that statistics should be used wisely. Many modern statistical tools apply probability theory to black-box modeling where causality is ignored and we are looking for the best fit with empirical data. The assumption of this application is that we are studying a modeled but true relationship, which has stochastic (random) variations that affect observed results. The relative weights of causality driven and random effects could be anywhere between causality only (1 vs. 0) and purely random (0 vs. 1). (Note: the relative weight 0.8 vs. 0.2 would indicate that 80% of observed results came for a reason and 20% from random fluctuations). If we have no random component, we do not need statistics—theoretically we have (almost) 100% prediction power. It may look obvious that we do not need statistics to send a rocket to the Moon, but this is not completely true because we have to keep in perspective all possible errors in measurements and actions. Thus, the $(1, 0)$ case exists only in pure mathematics. The other end of the spectrum (when we deal with purely random outcomes of infinitely big numbers of very simple and non-distinguishable elementary events) could also have almost 100% prediction power if we are interested in the overall picture, and not in the behavior of individual elements (where the prediction power is zero). For example, thermodynamics is built exactly this way. The results in the middle of the spectrum are much more complex, since they depend on valid separation of causality and randomness (which means they are model-dependent), but worse still—they depend on the number of events analyzed. Considering the number of important covariates, just creation of reliable models may require a huge number of trials, let alone testing research hypotheses in the final (even bigger) experiment.

The randomized controlled trials (RCTs) are the established hallmark of the evidence-based clinical research. However, RCTs employ some assumptions that should be clearly laid out. To make things simple, let us explore one case when the outcome is almost unequivocal dichotomy (e.g. 30-days mortality from any cause). The selected population of patients is assumed to have the true probability of dying within 30-days; all patients are equal in this respect and can be thought of as the same "biased" coin, with unknown (but existing) probability (usually not 0.5) of producing "heads" (death) instead of "tails" (survival). Next, suppose randomization guaranteed that two arms are absolutely balanced around this probability and both are ideal representations of all potential patients, who, in turn, are viewed as identical coins. Then it is assumed that the use of the drug shifts this unknown probability equally for all patients to whom the drug is given. After that, we use some advanced statistics to test the null hypothesis of the zero shift, compute the P-value, and if it is less than, say, 0.05, we declare the winner; otherwise, we announce a tie. Omitting all calculations we can say that the smaller this hypothetical difference is, the larger number of patients we should enroll. The question *which* difference has any clinical meaning is often a "judgment call", which in some cases means a randomly chosen number, and in other cases means an educated guess (or, as in some drug development programs, the judgement of a meaningful difference is based on commercial and regulatory guidance). Quoting the authors' favorite character: "It has long been known that one horse can run faster than another—but WHICH ONE? Differences are crucial." (Robert A. Heinlein "The Notebooks of Lazarus Long").

If we do not like any of these assumptions, we should perhaps lower our expectations of the predictive power of statistics. Let us make it clear that we are not calling a "Big Bluff"—statistics is the best tool available, and problem solving assumes discarding of unimportant noise and focusing on most important factors. This is what statistics is here for, but there is no guarantee that this always works. Statistics has certain limitations and some questions cannot be answered using statistics in principle, whereas some other questions cannot be answered from a practical point of view.

Let us conclude with a very simple example to demonstrate the proper and improper use of statistics. In this example, we would like to highlight a delicate difference between data-fitting models and model-driven experiments. In the former case, even with availability of large amounts of data, a researcher assumes limited knowledge of the studied mechanisms and simply seeks empirical evidence for explaining the phenomenon of interest. Frequently, efforts are made to provide a statistical description of surrounding noise in the data, instead of taking the position that the relationship may be mostly causality-driven, with some noise that should be not studied, but rather eliminated. In the latter case, the researcher seeks theory first and performs a theory-driven experiment to obtain supportive evidence by the data. The latter approach could be regarded by some as the model-based drug development or a *pharma-cometric* approach [34]. In the authors' opinion, the latter approach is much more scientifically sound than the former one. Of course, it may be not feasible in very complex settings (e.g. in immunology or neurology) where the true underlying mechanism of the disease is unknown and/or is very difficult to model. Yet, in cases when the real theory does exist, the data-driven modeling (that ignores our knowledge) is inappropriate, while the theory-driven approach (that explores this knowledge through experiment) is the way to go. Our example is as follows.

The task: Estimate change in total hemoglobin (THb) concentration after infusing 1 unit of red blood cells (RBC).

Improper approach: Run an observational study based on pre- and post-infusion readings (inclusion criteria: availability of data) using one or several big databases—the more data we have, the better. The result is one number with a confidence interval. In the discussion we include subgroup analyses around important (by our perception) covariates, limited by availability of the data: demographics, bleeding vs. non-bleeding, clinical settings, time of transfusion, baseline hemoglobin, etc. Note that the chances of using the most clinically relevant covariates are very low: the circulatory volume of patient and/or the volume and hemoglobin concentration of 1 unit of RBC will likely be unavailable. The value of the obtained results is minimal; inference is only limited to the population, not an individual patient. Future studies will explore the question in specific subgroups. It may take decades until critical covariates are elicited and the final list will not include all of them, but instead will include some that are absolutely irrelevant, since they demonstrated statistical significance. The fitted formula based on the analyzed data will be very complicated, impossible to use and plain wrong, but it will work, on average, for some subpopulations.

Proper approach: Apply theory (arithmetics of concentration in a mixture) and make some necessary assumptions (volume of a mixture is the sum of baseline volume and infusion volume). Derive a formula based on V (patient volume), v (infusion volume), Hb_p (patient hemoglobin) and Hb_s (solution hemoglobin). (The formula is $\Delta = (Hb_s - Hb_p)/(1 + V/v)$.) Conduct an observational or a prospectively designed study to compare prediction with observed results. Bleeding patients must be excluded. Additional fluid given between measurements should be accounted for, but it is better to exclude these cases as well. Data collection may be a challenge; V may have to be estimated (e.g. 5L×weight/70kg); parameters of 1 unit of RBC could vary greatly (both v and concentration Hb_s must be known as the inclusion criteria—otherwise we may have similar issues as in the "Improper Approach"). The comparison is done on the individual level (paired). Some variations are expected due to measurement errors; any serious deviations (outliers) are reviewed; any systematic shifts are analyzed. If the predicted and observed data agree well, then the task is complete. For now we confirmed the theory and have a prediction tool for every case. If not, we are back to reviewing theory and assumptions. For example, we may conduct a specific study to adjust an assumption that the total volume is $V + v$; maybe it is $V + kv$, where k is some specific "coefficient of dissolving", which is slightly lower than 1 (maybe 0.8). Then we go back to our data. We continue with such adjustments until we have agreement in data and prediction for *every case*.

1.6   In summary: what this book is all about

This book is neither a textbook, nor a systematic review, nor a collection of wise recommendations. This book was written with an intention to share summarized personal experiences of the authors, and, when possible, make generalizations that reflect the authors' opinions developed during many years of work in drug development. Some notes of caution: 1) These opinions are not necessarily consensual, which may lead to an impression that the book at some places is self-contradicting. (As we shall discuss, the adopted strategies may be highly dependent on the place of work and assigned duties; e.g. the same actions may lead to promotion in small pharma and to firing in large pharma, and vice versa.) 2) The opinions are likely disputable (as all opinions) and can look pretty controversial, since the main pre-condition of putting a lot of valuable time into this book was the absence of self-censoring.

Most of the examples come from the (still ongoing) two decades long history of development of an extremely problematic drug (hemoglobin-based oxygen-carrier) in the blood transfusion field that was developed before existing rules of drug development were established. Application of the pragmatic approach described in §1.2 permitted uncovering some major problems, mostly related to the theoretical inability to foresee everything, even with very well-thought out and established regulatory guidelines.

Before delving into the main chapters, let us add some clarifications.

The authors acknowledge that they cannot provide a "plain and clear" answer to the question of what is required for the biostatistician to maximize his or her value and impact. The right words would be: ensuring proper use of statistics, full engagement in design, execution and analyses of clinical development programs, maintaining coordination with all neighboring fields/departments, with all of the above aimed at one ultimate goal: the validity and effectiveness of the entire process. This would be a classical theoretical recommendation with minimal (if any) practical value, nothing to do with the pragmatic approach declared in the book title. The real answer, as usual, starts from "It depends". The short list of critical factors includes *seniority*, *credibility* and *place of work*. The first two factors are self-explanatory. In theory, the career ladder and credibility go in parallel, but in reality, credibility may either offset or amplify seniority. It is well known that keeping the perspective of the "big picture" is a prerequisite for moving up on both ladders. The winning strategy was described in an old fairy tale: two workers working side by side on a big construction site gave different answers to the question "What are you doing?" The first one was digging holes while the second one was building the city. While this perspective is crucial for a biostatistician at any level, what a person is actually asked to do is an important restrictive factor, which brings us to the place of work.

As mentioned in the preface, in this book we do not cover the work of biostatisticians in academia or regulatory agencies. All remaining work places could be roughly categorized into *big pharma* and *small pharma*. For our purpose, these terms have less to do with the size rather than with the practical distribution of work responsibilities. The biggest difference is high specialization in the big pharma compared to the small pharma. The title "biostatistician" has quite different meanings in these settings. For example, the first author, as a department head in a small pharma company, was responsible for data management, statistical programming, and biostatistics (including support for manufacturing, R&D, study design, study reports, BLA, and numerous publications). The second author, as a disease area lead statistician in a big pharma company, has more in-depth engagement in biostatistics activities (a much larger number of studies and projects within the company's portfolio), and is involved in statistical programming to a much lesser extent while interfacing data management activities only occasionally.

Furthermore, by our definition all CROs (regardless of their size) belong to the big pharma model. Consulting biostatisticians from academia, when they consult for big pharma, act based on the same model; when they consult for small pharma, it largely depends on a particular company, the expertise of the consultant, and relationships in this collaboration.

Some small pharma companies adopt the big pharma model by outsourcing their work to CROs—this may be very suboptimal in the long run. If a small pharma company does not have its own data management department, then, after years of clinical development (when data are pouring in from 10–15 different CROs), it may turn out that data reconciliation and establishment of ownership of all these data (let alone bringing it to the point of analysis-ready datasets) may be a very big challenge, and require efforts of a much larger magnitude than originally thought.

One may argue that high specialization and vertical separation of duties may potentially create a lot of problems. Modern business models (especially in the CRO settings) require that any task should be performed as cheaply as possible by those qualified to perform it. As a result, lack of understanding of what will be done at subsequent steps may lead to small oversights that necessarily pile up in big multi-step projects.

For instance, an increased specialization has led to situations when in many big CROs even statistical programmers who perform data cleaning and create raw datasets may barely talk with statistical programmers who create analysis datasets, let alone biostatisticians who develop the data specs. For the first author (during his recent contract work for a big CRO) it was an insane revelation that these 3 different functions work independently with minimal collaboration (after a decade of doing these tasks by himself in a small pharma company)! The irony of the situation is that fixing problems (when they arise) requires an over-qualified individual with hands-on knowledge of all steps in the process. Unfortunately, such a person does not belong to the existing business model: he or she is either long gone to another company, or beyond reach (maybe 3 levels higher) inside the company, and thus usually cannot be easily summoned. It looks like inside the existing model there is a need for an outside (highly-paid) professional problem solver who should be available on demand on short notice.

Many big companies declare that their goal is to prevent problems instead of solving them later. The only available tool to prevent problems inside a profit-oriented business model are well designed and strictly followed standard operating procedures (SOPs). On the one hand, this creates a system of roles and responsibilities which, in standard situations, works flawlessly as long as the appointed parties execute the tasks according to these SOPs. On the other hand, in non-standard, non-covered by SOPs situations, the final results can be disappointing; and yet the employees who executed the tasks are free from responsibility for these results because they followed all steps as prescribed.

In the authors' humble opinion, even the best-designed set of rules developed by truly knowledgeable experts cannot foresee everything—sooner or later there will be a situation when SOPs do not work and should be expanded to cover the non-standard situation. Note that there is an unintended, but rather naturally derived parallel with Gödel's incompleteness theorems from mathematical logic, which, in simple language, imply that no matter how hard you try to create a complete set of rules, the terminology that you use will permit/guarantee creation of a non-answerable question that will create a need for expansion of the system of axioms to answer this question (non-covered situation).

Overall, one of the biggest decisions in a biostatistician's career is to decide which working environment is preferable. The authors hope that this book will facilitate this decision for readers who have not made the definitive choice yet.

Introduction to Chapters 2, 3, and 4

An appropriately qualified and experienced statistician is a valuable player in any drug development team in the biopharmaceutical industry. This is well acknowledged and even articulated in the ICH E9 guidance [61]. The biostatistician is a *team* player, who interfaces and consults various multidisciplinary clinical development teams in the day-to-day work. To maximize impact, the biostatistician's skillset should be broad and diverse, and reflect technical, tactical, strategic, and regulatory requirements of drug development.

Recall that in §1.6 we mentioned one important distinction between small and big pharma models—namely, the scope of work responsibilities. In a small, one-drug pharma company it is common that a biostatistician is fully responsible for all kinds of data tasks, including statistical programming (SP), data management (DM), and biostatistics itself. Clearly, this calls for deep understanding and hands-on experience in all these aspects, including the design and organization of data collection, data cleaning/verification, data handling/manipulation, statistical programming and preparation of analysis datasets, and—as the last step—data analysis with subsequent interpretation of the results. It stands to reason that both DM and SP are major prerequisites for obtaining clean and reliable analysis-ready datasets that are passed to the biostatistician for statistical analysis.

In the big pharma model, there is high specialization and separation of duties, particularly in the field of biostatistics. From the authors' personal experience, this often creates discontinuity—since multiple people (and multiple organizations) are involved in the process (from developing the study protocol, organization and implementation of data collection, subsequent data cleaning and preparation of datasets), there is a risk that some small mistakes can pile up and lead to negative consequences in due course. Good understanding of the "big picture", including steps where things can go wrong, is essential.

Of course, we are not saying that any individual contributor biostatistician in the big pharma model must necessarily have the inside-out knowledge of all technical details related to DM and SP. Our main message here is as follows: Biostatistics is incomplete without DM and SP, and biostatisticians should definitely give big credit to their DM and SP colleagues, who perform often invisible, yet crucial tasks at the "back end".

In a sense, chapters 2, 3, and 4 describe a bridge between three fundamental technical pillars of drug development—statistical programming, data management, and biostatistics.

2

Statistical Programming

2.1 Introduction

As we have already mentioned, biostatisticians, regardless of their place of work, find themselves at an intersection of different fields (unless they deliberately restrict themselves to "pure" science). While it is almost impossible to name an area of research and development that is not influenced by biostatistics and vice versa, in reality, getting involved in all these areas is a long and sometimes painful process that usually runs in parallel with acquiring years and years of valuable experience.

Nevertheless, there are two questions every biostatistician faces immediately (and without exception) from the very first day of work in an attempt to produce any output:

- Which tools should I use?

- Which materials do I have to start with?

Evident inability to function without answering these two questions makes statistical programming (SP) and data management (DM) the true *Atlas* and *Caryatid* for biostatistics.

At the first glance, there is no issue. The entire history of humanity teaches us that progress is based on division of skills and delegation of responsibilities. Thus, let the data manager (or DM department) take care of data collection and supply me (the biostatistician) with data to analyze, and when I need something according to my (statistical) understanding of the situation, I will ask the statistical programmer (or SP department) to do it for me.

So, what is wrong in this picture? In theory, it should work flawlessly; moreover this is an existing business model employed by many big pharma companies, and every time it does not work, significant efforts are made to improve its functionality. In recent years it has become clear that division of skills and responsibilities necessitates proper coordination of activities across different fields. The words "coordination" and "integration" have practically become a "must be present" mantra in job descriptions and CVs and, as any mantra, it has almost lost its meaning.

As we already mentioned, it is not a question of who should coordinate these activities; it is the unique position of the biostatistician (who is forced to communicate with both the statistical programmer and the data manager) that precludes the answer.

The questions we want to discuss are essentially how far the biostatistician should go in crossing the lines while maintaining healthy pragmatism. The advice that would satisfy any first-grader: "You should become a better programmer than any programmer and a better data manager than any data manager" is obviously impractical. In what follows, we give a summary of very opinionated recommendations on what to learn and get involved in, and which investment of time and efforts (while may give you a high level of satisfaction) should rather be avoided.

We would like to emphasize that these recommendations are not designed as "rules to follow"; they are rather a demonstration of principles to be applied in different situations. For example, they highly depend on one's real work environment and job description and will dramatically differ for a staff biostatistician in a big CRO and for a biostatistician in a small pharma company (or even academia), where the biostatistician is, in fact, a statistical programmer and data manager at the same time.

2.2   Asking the right questions

"Eighteen creatures came to Answerer, neither walking nor flying, but simply appearing. Shivering in the cold glare of the stars, they gazed up at the massiveness of Answerer.
'If there is no distance', one asked, 'Then how can things be in other places?'
Answerer knew what the distance was, and what the places were. But he could not answer the question. There was distance, but not as these creatures saw it. And there were places, but in a different fashion from that which the creatures expected..."

Robert Sheckley "Ask a Foolish Question"

Let us start with the obvious: it is practically impossible to separate biostatistics from statistical programming. They interfere with and influence each other at so many levels that an all-inclusive list of such interactions would become a job description. The nature and the depth of this integration almost dictates the "weakest link" approach to this problem. A biostatistician who has no clear understanding of what statistical programming does is destined to fail. If he/she just wants to do it right and be limited to biostatistics work only, he/she will likely experience something similar to the following:

1.   Probably, the database is not ready for my planned analyses (unless some miracles happened at the previous stages of design, data col-

lection and creation of structured analysis datasets). Thus, I need a programmer to create analysis datasets for me.

2. If my analyses are complex enough, there might not be a straightforward procedure in my software (it is usually SAS for general analyses and some more specialized software for other analyses, but the reality is that SAS is almost always preferred for intermediate data handling). Thus, in order to perform statistical analyses, I need the programmer to get intermediate results and feed them properly into my next step(s).

3. Now that I have what I requested, it looks messy. Thus, to report it to the regulatory agency or to present it elsewhere (including internally) I need someone to do professional reporting.

Now, we shall add injury to assume that you have a good and experienced statistical programmer (of note, they are a hunted and endangered species—everyone wants them, but they are hard to find, especially with budget constraints—it will be shown later why this is the case). The programmer looks at the plan and asks the corresponding (clarifying) questions:

1. The dataset you are asking for is impossible to create: some of the data pieces were never collected (and were not supposed to be); some of the data were collected poorly (a lot of missing values, non-reliable); some of the data came from multiple sources, need conversions and adjustments; some of the data were collected outside of the normal settings of the study, and thus are present in a weird format and need cleaning with invocation of the source documents, let alone validation of the used software. I can propose some substitutes and finally create what you are asking for, but it is not as straightforward as you think and, of course, it will take more time and will cost more. Do you want to proceed?

2. If you insist, I can analyze data as per your request, but your proposed selection of procedures and their sequence seems to be very suboptimal—it can be done more efficiently and technically simpler. I have done similar work before, still have the set of macros that would do the job, but we need to go back and revise your design of analysis datasets (or add an extra step to create suitable input). What do you choose?

3. You probably understand that there is almost nothing in common between presentation and regulatory/production reporting. If you need production reporting then say so—I know the rules and I have a set of utility macros which, after necessary adjustments, could be used in our situation. If you need it for a colorful presentation, my advice would be: tell me what data you need, and I will transfer it into Excel where you can then create fancy graphs and then transfer them directly into Word or PowerPoint.

An additional note of caution from the programmer can follow that not everything may go as planned (especially for #2). Some strata may disappear; some results may be very different and may require serious adjustments to proceed to the next steps, etc. In short, you will need some serious additional exploratory analyses, and it is better to start planning for them outright.

It is clear that with such a (mythical) statistical programmer you have a good chance for success with your project and even more than that; you need him to succeed. The question is: does he need you?

Frankly, under this scenario, the biostatistician is the "weakest link". In 99% of the cases, the "unicorn" statistical programmer can do the entire project alone much more efficiently by adding to his knowledge some small internet research and/or a couple of hours of consultation with another statistician. No wonder that in a big CRO, a good biostatistician and an extremely proficient SAS programmer is usually one position. If he/she is really good, their career path will depend solely on small particulars but it generally progresses fast. Their positions can have different names, with no separation between biostatistics and statistical programming (e.g. manager, associate director, and finally, director). In other words, there is no big difference between a director of biostatistics and a director of statistical programming as an end point for individuals who come from this realm. Giving recommendations to this rare breed is not an intent of the current chapter—they do not need it, to begin with. We just want to acknowledge their existence and reasons behind the fact that they are so hard to find.

Instead, we shall focus on recommendations to biostatisticians who have minimal knowledge of statistical and presentation software, have no time to spare, but who want to become more or less independent from outside help, keeping as a distant goal proficiency in statistical programming.

The first author started his work as a statistical programmer/analyst in academia, got to the level of SAS advanced certified programmer, and then was teaching for several years a crash course, *SAS Programming in Clinical Trials*, in a computer science school. This experience, combined with subsequent work as a director of biostatistics, is the basis for recommendations that follow.

2.3   Choice of statistical and presentation software

The industry standard is SAS. Therefore, throughout this chapter, we focus solely on SAS. Our assumption is that the reader has some basic familiarity with SAS, e.g. through a short course from a University or some other training. Besides the fact that SAS is approved/accepted by all regulatory agencies, it is very convenient for people who like to have "all bases covered" while maintaining logical transparency. I (the first author) remember my first impression, which is probably typical for my generation and says little to young guys: "It is a database language that permits C programming in every row". Of course, there are software packages that would work better in some particular situations, but overall, in terms of data handling and abilities, SAS/STAT is the most pragmatic choice. The best part is that it will cover all your needs without spending a lot of time on learning how to work with multiple tools. In addition, it satisfies a good old 95/5 rule: "95% of what you will ever need, you can do after learning 5% of available information." The only place when I have found that "the game is not worth the candle" is using SAS for presentation purposes. For the latter, you should have a licensed SAS/GRAPH (which I did not have when I started working with SAS) and master some skills that are applicable mostly for the SAS environment. Having a flexibility to change something "on the fly" is important for presentations that are living documents (read: not finalized and subject to change from one presentation to another). Our recommendation is: create datasets in SAS and use Excel thereafter—this makes the entire Microsoft Office your friend for transferring and editing your graphs. Importantly, Excel provides not only basic statistics tools, but also some programming tools. It is not too complicated to create a simple macro in Excel as a base and then step up into Visual Basic code in order to achieve what you really need. We mention this just to make our first recommendation—do not waste your time on learning all the ins and outs of Excel—rather keep it as an auxiliary tool when you need some quick temporary fix. In case you do presentations for a living, it is better to invest some time to learn SAS/GRAPH and SAS Output Delivery System (ODS).

2.4 "95/5" rule

What are the magic 5% that guarantee your freedom and where can you find them with maximum efficiency? And what about the remaining 95%?

We shall try to separate and highlight topics that are absolutely necessary to learn and explain why they merit serious investment of time and efforts. Some of them are almost self-evident while others are often underestimated, overlooked and left for "better times". The common property is as follows: without them, one will be always crippled and will spend much more time in the long run. Unsurprisingly, most of the topics deemed "too hard" for the beginners are very easy to master after reaching some level of conceptual understanding (e.g. SAS Macro language). Strictly speaking, all the other stuff is optional—for many topics in SAS one should just have a general knowledge of what can be done and where to find help when (or if) it is needed.

In no way should these recommendations replace serious study of SAS, but they may help to decide where to start and, most importantly, where to stop. There are many good books on SAS programming, but they usually provide either too much or too little information relatively to one's particular needs.

The biggest challenge is creation of some minimal basic toolkit, which could be easily upgraded later on if your needs change for any reason. Let us make it clear that there are as many sets of necessary programming skills as the number of existing positions in biostatistics multiplied by the number of active biostatisticians. In other words, it defies "written in stone" rules in principle. It is quite possible that you would never use some of the provided recommendations and/or you would need to add something that you cannot live without which has not been even mentioned.

The ultimate purpose is to give an example (hopefully valuable) of the application of the pragmatic approach.

2.4.1   The sources

From the first author's experience, the most comprehensive crash course to start with is *SAS Certification Prep Guide (Base Programming)* from the SAS Institute. All editions and versions starting from the 1990's provide a great base for future work (and studying). Web-based versions are shorter, but the latest printed versions provide a more conceptual approach. Speaking of the conceptual approach, one of the most valuable manuals is *SAS Language Reference: Concepts* [91]. This book is an invaluable source of necessary information, especially if your knowledge of SAS is spotty and has never been organized systematically. We highly recommend skimming through this book and then using it as the best available help–chances are that everything that you need according to the "95/5" rule will be there. The next part, which is traditionally avoided in SAS Base courses, but is an absolute must, is *SAS Macro Language Reference* [92].

Recently, a similar guide for Advanced Programming became available and might be used to replace (or complement) these two books. Probably, this is a complete sufficient set of printed materials that one really needs. When the next version of SAS becomes available, you will be already using Help for all your future needs. If you work with clinical data, you will need, in addition, an introductory article on CDISC [28] and spend some time in browsing the CDISC website www.cdisc.org for official guides. This is it. Other great sources may be available, but at the beginning one may want to restrict the number of books on the shelf.

2.4.2   SAS Certification—Is it worth the time and efforts?

An "extremely original" answer is "It depends".

For Base Programming, in most cases the certification is the shortest and fastest way to build a foundation: when you are ready to start, you order the exam online in 2 months at the most convenient place near your home, and buy a manual. Essentially you burn all bridges, because either you invested about $300 or, even better, your company paid the bill, so you just have to pass this exam and thus you have motivation in place. Considering the fact that you already graduated and you are using SAS at your work place, if you really spend some time on preparation (at least 1 hour daily), it is almost impossible to fail. After 2 months you will have a nice SAS logo for your CV and some level of confidence.

For Advanced Programming, the common answer is "No", unless you plan to add a second occupation to biostatistics. The only exception is the situation when after 2–3 years (when you master all skills which we will discuss momentarily), you read the exam description and realize that you already know 90% (which means you were forced to use SAS extensively at your work) and are really curious about the remaining 10%. Then choose the closest SUGI or PHARMASUG conference and take the exam one day before. Keep in mind that this will take considerable efforts, and the positive outcome is not guaranteed. In the worst case, you will learn about your limitations, and in the best case you will get a respectable title of *Certified Advanced Programmer* which, together with your degree in biostatistics, will triple the number of managing positions available (you will have "statistical programming" in titles in addition to "biostatistics").

2.5   Data access, data creation and data storage

It is a nightmare for the beginners who get to a new place and start without any guidance—e.g. this can happen if you start working for a small company which decides that it is the time to get the data from multiple CROs under control and you are the first person to do this. Even a simple task of getting everything together and making it accessible from your SAS session requires basic, yet real understanding of how SAS files (or at least SAS libraries) are organized. You should be ready for the worst case scenario, which may include (but is not limited to):

- Separate databases for different clinical trials created by different SOPs (at least one per CRO).

- Absence[1] of Define.xml document.

- Absence of any analysis datasets.

- Absence of format libraries.

- Different SAS versions for different databases.

- Absence of any supporting programming package.

---

[1] "Absence" may mean either that your company does not have it or that nobody knows where it located, which is essentially the same.

- Absence of well-organized outputs (e.g. all results that you are supposed to verify exist only in study reports generated by a CRO).

If, on top of all this "pleasant" environment, your only experience in SAS was running examples from SAS Help or from an already created directory, you may be in real trouble. Your only ally is the IT person who installed SAS on your PC and gave access to some (specified by your manager) directories that contain the data of interest. Next are some recommendations on what can be done in this case.

1. Create a simple assignment program for every study with accurate LIBNAME definitions, including version number, for all folders that contain datasets for the entire study (Note: it could be more than one) and string OPTIONS NOFMTERR. Run the program and make sure that all assigned library names and every data set can be opened from SAS Explorer. Create a folder "(My) SAS programs" inside the "Study" folder, and save the program as "Assignment [Study name]" for each study. Take a deep breath—you have a rudimentary access to all available data.

2. Make full inventory of all available documentation for every study. Your first priority is formats, which can be in 2 forms: already created catalogs, or a program that creates them. Do it without regret for the time spent. Use an IT person to search for the "FORMAT" keyword through all directories, check all available SAS programs for this keyword, check whether you have a study annotated CRF (this will permit you to figure formats by yourself and write the appropriate code), and, as a last resort, contact the CROs. Do not stop until you have them all. Keep in mind that without formats you do not have your data.

   When contacting the CROs, try to get from them the entire programming package (make sure that it is executable, meaning that it contains all utilities macros that were used to generate the results and the analytical database) and other documentation (Define.xml, annotated CRF, ReadMe file, and the created output including tables and graphs) for each study. Some possible obstacles are: 1) the CRO or a department of the CRO that generated the package does not exist anymore; 2) the contract with the CRO did not allow for the programming package or even the created output. For 2), you can still require the output, but the programming package could be declared the Intellectual Property (IP) of the CRO. If you plan on *owning the data*, every effort should be made to get all this stuff, starting from the official warranty to never use this code outside of the study and including paying extra money for this IP. For 1), the situation depends on what documentation your company has. Essentially, in the worst case scenario you may end up with the study report and the database—this will preclude or extremely complicate any additional analyses of the study, unless you are able to recreate any related numbers from the study report. Chances are that you will have a choice between using only the already reported results or disregard them completely and re-analyze the study.

3. Organize the data and documentation properly. Think it through carefully while choosing the organizational structure—you will be working with whatever you create for a long time. For example, keep in mind a possibility of working with two or more studies in one session, which may require (especially if you have data in different places) using the %INCLUDE statement for assignments with different names or libraries for different studies.

4. What will you need to learn in SAS either at the beginning or along the way? Surprisingly, not much: LIBNAME and FILENAME statements, %INCLUDE statement, options NOFMTERR and FMTSEARCH for user-defined formats, PROC FORMAT to create formats, basic knowledge of SAS-defined formats, and major date/time formats (chances are you will not need others).

Let us explain what is meant by "owning the data" for the study, which will explain why all steps described above are so important. The assumption is that you will have:

- The entire collected data (cleaned and verified against the source documents that are available as well).

- All generated (and already reported) results together with the programs that produced them.

- An ability to run new analyses based on strong confidence in the quality of the collected data and all already generated results, or a possibility to make properly documented corrections if new analyses identify some errors.

As one can see, this work (that is typically done by a SAS programmer) involves much more data management than programming, without any hint of statistics.

2.6   Getting data from external files

This is another situation when you may have to choose between losing independence or making relatively small investments of time. Chances are that sooner or later you will face the necessity to use data that was collected outside of traditional SAS-based sources: it could be additional data for clinical trial (e.g. lab norms, extensive follow-up that went beyond study design, observational data for historical controls, etc.), data from census or additional health economics data, data from related pre-clinical studies that you would need to use and that is rarely collected in SAS, the results of analysis generated by different software packages, etc. The format could be different, but in 99% of the cases it will be some kind of text-based files or Excel (or rarely even Access) files.

For most text-based files, which are normally organized horizontally, the usual read-in is done using CARDS or DATALINES statements after simple copy/paste into the code. The biggest pro is that the syntax of this data read is quite simple, because you can restrict yourself to reading only rectangular data (say, N rows with m variables, which could be either character or numeric) based on their lengths or chosen delimiter. The biggest con is the necessity to keep original documentation and code for reading with DATALINES, and have a procedure in place to make sure that your code is in agreement with the source documentation.

A better alternative is the direct reference to the file by location/name using proper extension. In this case, you only need to verify the code (and there is no hardcoding involved). Surprisingly, it is usually both more reliable and does not involve extra knowledge beyond referencing the file itself. Moreover, there is one possibility that impressed me nearly 15 years ago. I had to read a file that I could not open because it was created inside some software that I did not have. To my delight, this was not a problem for SAS (after using the original file extension). The only trick that I used is worth mentioning (because I had no idea how data were organized): I assigned the only variable in the dataset to an automatic variable _INFILE_ which in SAS is extremely long and contains the entire row from the file that was processed. From here you can go by either: a) creating a text file using PROC PRINTTO and then reading it properly, or b) using the SUBSTR function to extract the data from a long row.

For all other needs (Excel, Access, JMP, SPSS, etc.), the only thing that you would need is PROC IMPORT (or PROC EXPORT), which has quite a simple syntax that can be bypassed by using interactive screens in SAS Explorer.

In addition to the already mentioned functions and statements, you have to be/get really familiar with PUT and INPUT functions and statements.

One important word of advice: there are a lot of tools and options available in SAS to do it at the highest level. Do not spend time on them outright; do it only if you are facing a lot of repetitions. After all, the only thing that you need is to get data into SAS; everything else can be done inside DATA step(s).

A useful application of PROC IMPORT and PROC EXPORT will be described later in Chapter 3 for the data cleaning project that involved work of multiple monitors for source verification of already existing lab data with essential data flow as follows: existing SAS dataset $\rightarrow$ (PROC EXPORT) $\rightarrow$

Excel Table → (corrections) → new Excel Table → (PROC IMPORT) → new SAS dataset.

2.7   Data handling

2.7.1   The DATA step

The DATA step is a powerful tool that permits almost everything that you need with minimal knowledge. When a new dataset is created from existing datasets, you may just consider a simple SET statement (preferably accompanied with the BY statement, especially when combining multiple datasets, which saves you at least one sorting) and the MERGE statement with a mandatory BY statement (preferably one to one, or, at most, one to many). Of course, multiple nested SET statements, exotic MERGE and FETCH statements can do wonders, but chances are that in the long run you may not need them.

The list of the needed options is pretty modest as well. In order to control the number of variables for input, use KEEP= and DROP= at the beginning, and to control the number of variables in the output, use KEEP and DROP at the end. It is worth spending time on understanding the sequence of execution related to RENAME= at the beginning and RENAME at the end (this may not be intuitive and might require some practice).

Proper usage of IF and WHERE to restrict the number of observations can be even more confusing. Our recommendation is to use WHERE= only for the input and IF (or IF THEN OUTPUT) construction for the output. The benefit of using WHERE= is two-fold: it restricts the number of processed observations, and it permits running of basic statistical procedures on subsets of interest. The use of FORMAT and LABEL assignments is recommended for the output variables.

There are two topics that are often overlooked. These may create enormous difficulties for writing even a relatively simple code. We insist on investing time and efforts in studying them in detail—this will save a lot of time and help avoid frustrations later:

- Program Data Vector (PDV), which is a collection of all variables that you have in a processed row, including all read-in variables, created (intentionally or not) user-defined variables (e.g. counters in the loops).

- Automatic variables created by SAS depending on used statements.

Generally, all your coding inside the DATA step is aimed at some operations with the PDV. You have to be in full control of this environment and clearly understand what you start with, which value you would expect for every variable used in your code at any moment, and what part of this PDV would make it to the output dataset (or multiple datasets) according to your design.

What you start with: variables with their values that you brought from existing datasets. The values of created variables are usually missing unless you retain them using a statement with the same name or they were created to indicate whether a record had originated from one of the input datasets by using IN=NAME for such indicators. The most useful automatic variables are: 1) _N_, which is a number of the processed PDV; 2) created using the BY statement FIRST.NAME and LAST.NAME for each variable that participated in the BY statement; and 3) END=LASTOBS that creates an indicator for the last processed observation.

The only advice that can be given is to follow logic in your code and remember that any changes in values will occur only because you asked SAS to do it or go to the next PDV. For any calculations that involve multiple PDVs, it is important to decide in advance on the following three things: when you start (which usually means that you should assign the set of initial values); what you want to carry to the next PDV using retained variables; and when you finish processing the group that is a subject of calculations. Theoretically it is possible to make reassignment on the last processed PDV after you are done with the output, but in practice it is easier to do a record of the group of interest at first.

Decide upfront whether you plan to have one output per PDV, multiple outputs (possibly to different datasets) per PDV, or one output per group of PDVs. A general recommendation is to use clear output statements for any cases with the exception of a natural output in a single dataset with one record per PDV. Code your actions based on auxiliary variables created using the BY statement.

2.7.2   Loops and arrays

These tools are a must. Arrays usually permit us to avoid boring typing when similar actions are performed with a big set of variables. The simplest application involves creation of a one-dimensional array with a fixed number of variables that are numbered, and processing the array with a basic loop DO I=1 TO N. A general recommendation is to name properly 3 arrays of the same length: first one for the processed variables, second one for the retained (carryover) values, and third one for the output. It may look like a waste of time, but in practice it streamlines logic and simplifies code significantly. This technique can be used for finding totals, minimums, maximums, running averages etc. (Note: number 3 for arrays is minimal—you may need many more carryover variables and/or output variables.) Often times these tasks could be done using PROC FREQ, PROC MEANS, or PROC SORT, but with a big number of variables the code will be much simpler technically with the array processing; the alternatives usually involve PROC TRANSPOSE on top of macro processing with creation of multiple intermediate datasets.

Note that *multidimensional* arrays will require nested loops, which is not complicated, but rarely needed.

One extra use of loops (and sometimes arrays) is generation of datasets (e.g. for simulation) when the OUTPUT statement is located inside the loop. An interesting detail is that even when producing thousands of records in the output dataset, you may use a single PDV for the entire DATA step.

2.7.3   Going from vertical to horizontal datasets and vice versa

This is probably the most needed and the most performed operation with clinical data. One common example is the lab data collected by visit with one record per visit with multiple assays per visit (horizontal) or multiple records per visit and a single record per assay (vertical). Using PROC TRANSPOSE is complicated by variables that are common per visit and by the fact that there are multiple variables (sometimes even different types) per assay to describe the results. The most practical approach in both directions is to use multiple arrays with length equal to the number of assays. For the first task, we are splitting existing PDV into multiple records without touching visit info with one output statement per assay (array element). For the second task, we have to create arrays of interest and retain them during processing of a group of records related to the visit so that an output is made only per such groups of records. As was discussed before, it is recommended to set all arrays to missing for the first record related to the visit, and have an OUTPUT statement for the last one. The technique is simple and straightforward, but should be mastered first.

2.8   Why do we need basic knowledge of the Macro language?

The major reason is that macro variables are the simplest vehicle to pass info across the program. As we noted before, this problem inside the DATA step is solved by creation of temporary retained variables, but when the DATA step is complete, this information is either lost or becomes difficult to retrieve. For example, accessing this information in open code is impossible, and getting this info back requires an extra DATA step. To make things worse, this is one of the most complicated techniques in SAS unless you use macro variables. This task is solved easily with the CALL SYMPUT routine. We would recommend using just 3 simple techniques for creation of macro variables, but you should understand them deeply:

- CALL SYMPUT inside DATA step.

- %LET in the open code.

- Arguments in invoked macros.

"Deep understanding" essentially means knowledge about visibility scope (where you will be able to access macro variables) and how to get them. It is absolutely necessary to be proficient in usage of multiple "&" signs and how to resolve them—they are mostly required if you derive a name from two or more macro variables (note that some of them could be counters in your loops).

2.8.1   Open code vs. DATA step

One of the problems in SAS programming is the difference in what you can do in the DATA step or in the open code (between data steps). We propose a very pragmatic solution, assuming you are already satisfied with what you can do in the DATA step. Most of your knowledge is easily transferable to the open code either by using the "%" sign for all statements or using invocation of the %SYSFUNC. Another valuable opportunity is the DATA _NULL_ step that does not create any data set, but shares all the advantages of a simpler syntax inside the DATA step.

2.8.2   Loops in the open code (inside macros) and nested macros

These tools are dreadful for most beginners and, on many occasions, are rightfully so. In short, they do not require advanced programming skills, but rather an advanced logic. When used properly, they can save a lot of time and typing, but it requires serious discipline and organization, let alone firm knowledge of the basics and thorough debugging. Thus, do not start from them, but if you feel that you are ready, do not hesitate adding them into your arsenal. Example of an application:

1. You worked out (and debugged) some useful macro.

2. You realize that you could reuse it with a small modification for the next step.

3. It is clear that adding another argument can solve many of your problems.

4. You want to run this macro several times with different arguments sequentially.

5. You create an outer macro, which is essentially a loop that invokes your original, but modified macro with a different sets of arguments.

6. Your code consists of the modified macro and the consolidating macro (loop), where both of them could be adjusted later for other (similar) problems.

Note: You may get tempted to disassemble any macro in repeating blocks, which became lower level nested inside macros. You may get a better presentation of your code, but you may start losing track of the logic. (It is the way of the utilities macro, which we will address later.) Thus, keep the balance.

2.8.3 Use of pre-written (by others) macro code

Sure, we have to do this—after all, nobody wants to reinvent the wheel. There is one condition, though—you have to understand any program that you use— at least what it does and, most importantly, how. The code written by others is the best educational source that you can have, especially when you need to use it, which is the best possible motivation. Of course you should draw the line somewhere—you obviously cannot afford to study how SAS procedures work at the low-level, and the same applies to some pre-written macros that are about to become standard procedures in SAS/STAT (a rule of thumb for recognizing this is a huge number of (silent) arguments or controlling options in the description). Critical for your education are short technical macros (e.g. assignments, reporting, formatting, etc.). Even if you do not need a technique now, you will learn about the existence of some possibilities, which will simplify your life in the long term.

Another reason is avoidance of being an "algorithmic guy" from an old joke about boiling water using the best and most complete algorithm. (In short, the full algorithm: take the kettle from the shelf, open the faucet, fill out the kettle, close the faucet, and put the kettle on the electrical stove; start warming, stop after done. The guy already has a full kettle on the stove, but goes by creating the pre-condition for the algorithm by emptying the kettle and putting it back on the shelf. Then he runs the full algorithm again.)

Jokes aside, I (the first author) have seen it many times. Once, working as a programmer, I got a cry for help from my colleague biostatistician, who was using some powerful and tested macro from a statistical journal to analyze some data. At the time SAS was executing this program without any error message for 4 days without any visible results. When I stepped inside the code, I realized that a) 99% of the work that this macro was doing was unnecessary for getting the results of interest (for my colleague) and b) the data was too big for the PC that was running the macro and results should be expected within a week or two. Considering the fact that the intended analysis was exploratory and it was planned to be done repeatedly, waiting for the results was not a good option. Within the next 10 minutes I bypassed unnecessary blocks, started the program and went outside for a short smoking break. When I got back to my office I found a note: It is working! Already done! The year was 1999 and personal computers were not as powerful as today, but some algorithms still require significant times for execution.

2.9   Summary

In this chapter we have discussed some important aspects of SAS statistical programming in clinical trial research. As already mentioned, our recommendations here are not written in stone; they rather provide incremental steps (note: in no particular order) for self-improvement. A "guiding principle" for the proper use of these recommendations can be formulated as follows: If, after reading a specific topic, you have no idea what it means and how it can help you, then you are not prepared for this step yet. You can revisit it later, with improved knowledge and/or increased demand for something that you are not able to meet. We do guarantee that at the point when you master everything that has been mentioned (or even earlier), you would not need any future guidance.

3

Data Management

"We are what we process and what we do."

Colloquial wisdom.

3.1 Introduction

The true value of well organized and well performed data management (DM) is extremely hard to overestimate. In order to have both reliable and useful data for analysis, many steps in an inherently iterative process are to be made. We would like to mention outright that by "data management" we mean not just maintenance of clinical databases (e.g. ensuring that the data are stored correctly, its integrity is maintained, and everything that is needed can be easily retrieved), but a much broader scope, including database design, implementation, cleaning, etc. Let us try to list these steps and fill in the meanings of words that do not tell us much by themselves.

I.   Design of data collection (§3.2).

II. Organization of data collection (§3.3).

III. Data collection itself (we do not discuss it in this book because it is normally beyond the scope of a biostatistician's work; it is typically performed by study monitors or other technical staff).

IV. Data cleaning or verification (§3.4).

V.   Re-structuring of the data (§3.5).

Upon completion of these steps, the DM is done and the data should be ready for analysis and reporting.

While these steps look like a complete recipe for success, there are some meta issues, namely: "What is the big goal?" and "Who is doing what?" The most important issue is whether a biostatistician, who is performing data analysis and reporting, will be involved in all of the aforementioned steps or not.

Let us again distinguish two major settings: the big pharma model, and the small pharma model or academic research. The difference between these settings is, first of all, in the level of engagement of a biostatistician in different steps of the process. It would be fair to say that in the big pharma model, different "data tasks" will generally be split among several individuals: the trial biostatistician, the statistical programmer, and the data manager. The biostatistician would generally be responsible for selecting statistical methodologies, and writing statistical sections of the study protocol, developing a statistical analysis plan (SAP), writing/managing data specs, coordinating statistical analysis/reporting activities, checking the statistical output (which may be produced according to the SAP by the statistical programmer), etc. Just by the nature of this operating model, the trial biostatistician may not be deeply involved in all DM and/or statistical programming activities. In a big CRO, even this level of continuity may not exist; the duties described above could be divided between (any!) number of biostatisticians. A consequence of the technological revolution (division of skills and responsibilities) is a double-edged sword: a potential for increased efficiency (more studies/projects can be accomplished in less time) should be weighed against the risk that things can go wrong if the process is not properly coordinated.

In academic research, the situation can be very different. Here is an example of the *extreme* case: the biostatistician is a member of a small research team; he/she is responsible for most of the data tasks: helps the medical investigator to design a study; helps the data manager (or even just the study monitor who extracts clinical data) to collect data and/or access existing databases; participates in cleaning/verification activities; structures all collected/extracted data; performs statistical analysis and prepares the results for reporting; and even actively participates in writing manuscript(s) for scientific journal(s). Such a biostatistician is just short of being a magician. The list of necessary skills starts from the deep knowledge of the disease area, includes DM, programming, project management, and ends up with the medical writing. On top of this, any pre-written SOPs may be absent—all this may require a lot of creativity and good problem solving skills. In this chapter we shall try to describe only the DM skills that are (or could be) needed.

In small pharma, as in academic research, the extreme case described above is quite possible (the first author was actually forced into such a situation), but generally there could be a spectrum of possibilities between the "extreme case" and "minimally sufficient involvement". There are a couple of aspects that could make things for an aspiring biostatistician in small pharma even more challenging than in academic research. The first issue (which is also common to big pharma) is very rigorous requirements for data collection, analysis and reporting, imposed by the regulatory agencies. The second issue is an incredibly entertaining opportunity to fix some problematic work done by others (and deemed "impossible to fix"), while using relatively modest resources. In this chapter we shall present two case studies (§3.6–3.7) illustrating these ideas.

3.2 Design of data collection

"If you want something done right, do it yourself."

Charles-Guillaume Étienne

We start with a theory of how data management should operate from a biostatistician's perspective. It looks simple, but in fact it starts with the following questions:

- Why do we need the data?

- What kind of data do we need?

The answer to the first question is two-fold: to learn/study something, and to report the results of this study/learning process. These two parts actually split an answer to the second question into two parts as well: the first part is more or less creative, while the second part is largely mandatory and is defined either by the requirements from the regulatory agencies in the pharma industry or the requirements of scientific journals in academia.

Let us look first at what lies inside the "creative" part. It is assumed that we know the question(s) we want to answer by running our study; we have chosen the tools (some statistical procedures); and we now want to describe the data needed to generate the answers. Usually this is done in the SAP based on the study protocol, or it actually may be the opposite—we may write the study protocol according to the already prepared SAP. The obvious requirement—they should be in agreement. In reality, this is never easy—we may have two documents that are edited simultaneously, and there is a challenge to keep them aligned. Another issue is the fact that both documents are usually self-repetitive in different parts, and it is again not simple to remember which parts contain the same info, especially when several people are editing both documents. We would need a single person coordinating the efforts and one may argue that the biostatistician could be a good candidate.

Now we can involve the data manager (which could be just another "hat" of the biostatistician). An important question is "Can the required data be collected"? This is a very different question for the data extraction from existing documents or databases and for the data collection in a prospectively designed study. In the former case, if the answer is "No", then we are back to the drawing board looking for a substitute that permits us to give a "Yes" answer. In the latter case, the above question is frequently not asked, which leads to the problem of missing values and even sometimes inability to do scientifically sound statistical analysis. The red flag is that the data in question are not routinely collected. In most cases, this means just making extra efforts to ensure these data are collected, but sometimes the true question is whether the goal is realistic or not. For example, in a non-planned medical emergency setting (such as trauma, heart attack, cardiac arrest, etc.) it is unwise to expect that the data that is not usually collected will be collected, because it requires a change of the existing well-established protocol. Another example is requesting multiple significant (say, hourly, by 100 ml) blood samples in a setting of blood loss—we will end up with non-collected data. In any case, the section on "handling of missing values" should be present in any protocol and SAP.

The "mandatory" part (a subset of normally collected data that will be needed for a designed study) is actually easier—while it can vary in different settings, it is usually a routine task. Furthermore, this data is not critical to study design and it rarely represents any problem for collection.

3.3 Organization of data collection

In essence, this is a process of providing means to get the required data into our possession. This process is a cornerstone of data management. By its nature, this process is extremely complex and convoluted; it includes many different sources and routes; and it requires clear understanding of the final database structure. On top of this, it is subject to multiple rules and regulations.

Note: We have already (silently) restricted the meaning of the terms "data" and "database" to "data used in analyses". E.g. for an RCT we are not discussing the collection of CVs of principal investigators, site training records, etc.

The problems in the formal description of the data flow start from the very beginning, because we always have a combination of "data extraction" and "data collection". Let us start from the definition of these two terms. The obvious distinction between "already existing" and "newly collected" data is useless from the standpoint of modern data management. A critical difference is actually in the *format* of the data.

If some previously collected data are already in a structured electronic format that permits data manipulation programmatically, we can call it "already existing" and will use the term "data extraction"; otherwise we should organize this data in a proper format and this process will have all the attributes of "data collection". For example, processing of the previously collected medical records of patients from an observational study (if they are not in the described format), will be no different from actual data collection for an ongoing study.

In a sense, *data collection* could be defined as organizing data (doesn't matter old or new) into an electronic format that permits data manipulation, and *data extraction* is re-organizing the data into a pre-designed structure, which contains all analyzable data that is relatively easy to clean or verify (not necessarily to analyze!).

The entire design should include:

1. Evaluation of multiple sources of the data (which may also include securing data access) such as medical records or an electronic database.

2. A thorough design of the structure to keep all needed data.

3. A vehicle to get these data from multiple sources into a pre-designed structure.

The vehicle is usually some sort of a programming code for the extraction and (now it is usually electronic) case report form (CRF) or its analog, mapped to a pre-designated structure for the collection.

A typical example for an RCT: the main vehicle is an annotated CRF mapped to the SDTM for the collected data (there could be some intermediate databases, but we will not go too deep). For the lab data or some part of it (specific measurements, or central labs with an electronic database) it could be an extraction mapped to the SDTM. It will be certainly an extraction for lab norms, and with multiple labs it will be multiple extractions.

Of note, keeping the biostatistician's perspective, we are not discussing physical aspects of organization of data collection; instead, we are focusing solely on intellectual aspects of this process.

Since organization of data collection is Step II (according to our definition in §3.1), starting from here we have an iterative process at two different levels: inside Step II and going back to Step I. Even in a strictly regulated pharma environment (where every small task is the subject of a well-developed system of SOPs), it is impossible to get everything right from the first attempt. In real life, even with flawless execution, one is likely to encounter some issues that are either not covered by the SOPs at all, or the existing SOPs may be ill-suited for solving the task at hand. Therefore, going back to Step I and changing some elements of the design and execution are not necessarily an indication that some errors were made; it may be just a consequence of the complex nature of the entire process. In some sense, it is much easier for an experienced coordinator to get everything right with fewer iterations in a less regulated environment.

One problem in a strictly organized environment is that a person who is doing a specific work and has developed (by experience) deep understanding of the tasks at hand, may lack the authority to change the rules which were developed by higher positioned people who tried to foresee all possible situations.

In situations when the rules do not apply, there are two possibilities: changing the rules, or trying to squeeze the situation into the existing rules. In today's business environment, there is usually lack of time (or resources) to do the first and the second usually creates more problems than it solves. Strictly speaking, in practice such situations often go unresolved with a chance to be resolved later by creation of a new specialized SOP (or exception from the existing rule) that is suited to such a situation. There is no guarantee though that this SOP will ever be needed again.

A widespread implementation of the CDISC, while being very positive overall, shares the responsibility of creating such situations as well.

3.4   Data cleaning or verification

Speaking of making and fixing mistakes—this step is an inevitable part of any data collection process. Theoretically, a well-written system of well-thought edit checks can catch any possible problem of data collection (and data entry) and fix it after resolving all generated queries by a direct data verification against the used source. Such a system is usually designed by clinical programmers who have deep knowledge of a particular data capture system and its possible flaws.

We agree that most mistakes of data entry can be caught and fixed this way, when a residual rate of errors depends mostly on the time and effort that one is willing to spend. For example, the second independent data entry has a good chance to catch a very high percentage of errors, and well qualified personnel can review such cases and fix all errors that have been picked up. Triple data entry is even better, but it costs money and time. A usual tool to decide whether to go this expensive and time consuming route (after initial cleaning was done) is a random audit that defines error rate for each dataset and permits us to declare it "clean and sound" if the pre-defined standard was met, or continue cleaning this dataset, either by creating new edit checks based on errors that were discovered, or (if you get desperate) make an independent data entry again. This process should take care of data entry mistakes and converges ideally either from the first attempt (if you have a very experienced team designing the system of edit checks) or after one or two iterations (when some extra edit checks have been added).

In practice, the situation is much more complex. The biggest problem that cannot be resolved this way is the data collected wrongly. Let us look at several examples.

- In the demographic dataset, for some patient we have caught weight=842 kg. The monitor who is resolving the generated query finds that the source document (which could be anything from a medical record to eCRF) has this very number reported. The query is resolved, and the data entry is confirmed. The question is what to do next. If for this particular patient this info was collected only once, we can set it to missing (please let us not forget filling out a lot of papers because we are potentially "tampering with the data"!) The next question is whether (and how) we draw the line to declare this data incorrect? 400kg? or 300kg? Is it possible to foresee and address all such situations?

- Consider a similar situation, when the total hemoglobin (THb)=42 g/dL. A theoretically high value is 18g/dL, which is Lance Armstrong's after forbidden erythropoietin (EPO) and training in high altitude for months. Sure, we should throw away anything higher than 20 g/dL, and, more importantly, we can do it (and, actually, must). Such dramatic errors for electronically filled CRF could be avoided by imposing some permissible boundary for data entry; however it will not solve the type of problems we discuss next. For instance, what would we do with 15 g/dL for a 70-year old enrolled in a study of acute severe anemia, if this is one of the numbers in the sequence 6, 7, 8, 7.5, 15, 7.8, 8.2 and we plan on analyzing AUC for THb over time? We can make the case that "15" is the wrong number (even if it is confirmed by the monitor, and again, writing a lot of supportive documentation) and set it to missing. This is relatively simple, even though it may be time consuming (if, say, this patient is part of a critical efficacy analysis of a pivotal phase III study). But what should we do with the values "6" and "8" in the above sequence (even though the trial protocol did not plan for any transfusion between "7.5" and "7.8" and there was no reported transfusion or blood loss between the neighboring visits)? And, by the way, what are our chances to catch this using edit checks?

The list of such situations with no satisfactory resolution can be continued. And it is not so rare—according to some estimates, near 5% of data collected in emergency rooms is simply incorrect. There is no way to fix it; sometimes it is neither possible to drop it nor to use it, and of course, such cases might be extremely difficult to catch.

Another big potential problem is a complicated data flow, where you can even find a data entry error but cannot fix it without very deep understanding of the data structure. This situation will be reviewed in detail in the second case study (§3.7).

One final remark on a very useful and practical feature of data collection: the *database lock*. It is done when data are declared cleaned at the end of the discussed step. This is great, but as we can see, some errors can be found only during the analysis or reporting stages! Even if we know what to do, unlocking the database and locking it again after fixing the issue(s) can be impractical. One popular tool in these situations is ERRATA, which is essentially a SAS macro code that applies all fixes before analysis or reporting of the collected data.

3.5   Re-structuring of the data

The data that can be easily collected and cleaned/verified may not be convenient for complicated analyses. This is when re-structuring of the cleaned data is done. This work is usually performed by the statistical programmer, as we discussed in Chapter 2. In the modern pharma environment, a combination of the data specs and the programming code based on these data specs transforms SDTM into ADAM. One potential problem here is mostly in discrepancies inside ADAM recommendations, different self-contradicting SOPs, and differences between ADAM rules and their implementations in the SOPs.

Before CDISC, the potential problem was in combining data from different sources to create structured databases (SDBs) if data flow was complex enough; see the second case study (§3.7). In the academic research, we have practically a pre-CDISC environment when SDBs or analytical databases should be designed first and then created.

We shall now discuss two examples of data cleaning. The first one is for the data extracted from a huge Veterans Affairs (VA) database (in the academic research setting), and the second one is 100% verification of an enormous lab database in a pivotal phase III study (in the small pharma setting); this database was previously subject to multiple unsuccessful attempts to clean it.

3.6   First case study

The purpose of this case study is to demonstrate that we should be careful while making "common sense" decisions, because some (initially omitted) important information can change the "obvious" solution. We should look for this important information—always. The described problem was one of the many challenges encountered while conducting a long several-year grant funded by VA.

Background: We have patient/visit data extracted from a huge VA database, with one record per patient per visit, and we are interested in assigning *gender* for each patient in our own demographic dataset (one record per patient) with about 60,000 records. The extracted visit data has, on average, 10 visits per patient.

The problem: On top of that 2–3% patients have their gender information missing, for about 5% of the patients we have self-contradicting data when two different genders are reported for a patient. We already decided how to proceed with non-self-contradicting discrepancies; e.g. when some records have gender (but only one) and some do not, which were encountered in about 10–15% of patients. Since gender was a critical covariate in the derived logistic regression model, we can tolerate 2–3% missing data, but not 7–8%. Of course, there was no way to write queries to each of the 22 VA facilities across the US to resolve discrepancies; thus we had to make the rules for an automatic (programmed) assignment for the gender. There were two "obvious" solutions that are commonly used in such cases: a) use the last record, and if it is missing, go by the majority and use the last non-missing record as a tie-breaker; b) since our subset of data was representing just data for 3 years, the last record loses part of the value (the significance of the very last fixed value); thus we can go by the majority and use the last non-missing record as a tie-breaker. At first, we were inclined to use a), but after realizing that half of the patients changed gender over time more than once, we actually almost decided to go with b). It was imperfect, but, it should have provided better results than any other imputation technique.

The solution: Since we also had some other issues with data reporting, we arranged several meetings: a conference call with a programmer at Austin TX (where the mainframe SAS database was located), and a call with a front desk employee in one of the facilities (which happened to be across the street). During these meetings we learned something that completely changed our preferred solution. The gender information at each facility was entered once—during the first visit, irrespective of the already reported visits at different facilities, and then it was retained automatically. If there is a need to make correction of the first entry, it is done at the time when the problem is discovered, and then it is retained automatically. A quick check of our database revealed that inside every facility, the gender was stable for the majority of patients, and when it changed, it changed just once.

After we assigned the last record per patient/facility as the gender reported for the patient at this facility, most of self-contradictions disappeared, but we still had issues for 1–2% of the patients. Subsequent rules were, in fact, imputations based on common sense: the gender from most facilities was assigned for the patients who had different gender information at different facilities (the assumption is that reporting is correct more likely than not); as a tiebreaker (usually for 2 facilities) we used the value of the gender from the facility with the most visits. (Note: we first tried to use gender from the facility when it was changed, but it was an empty rule—all such cases were already resolved). After applying a SAS macro for the gender assignment, we found that we still had 2 or 3 patients for whom we could not assign gender using anything better than flipping a coin. In such cases we assigned the gender value to missing.

A summary of cleaning/imputation (by memory, since the author has no access to this data anymore):

- Patients in question—near 300.

- Resolved after applying the "inside facility" rule—near 240.

- Resolved after applying the "majority of facilities" rule—near 30 (usually one or two facilities with a big number of visits, and the rest with a much smaller number of visits, e.g. 1–2).

- Resolved for 2 facilities (there were no cases with 4, 6, etc.) by choosing the facility with more records—near 25 (in about 20 cases the difference between the facilities in the number of visits was big; in 5 cases they were close).

- Non-resolved—near 5.

Looking back, the real number of non-resolved cases is about 10. According to the front desk person, these are most likely the people who spent their summer in one place and winter in another place, with a comparable number of visits at two different facilities.

Major lesson learned: Knowledge of the rules used for data collection and data processing is a must for data cleaning and imputation. Without it, common sense can be misleading.

3.7   Second case study

"To consult the statistician after an experiment is finished is often merely to ask him to conduct a post mortem examination. He can perhaps say what the experiment died of."

R. A. Fisher

Foreword: This case study actually illustrates numerous points, starting from the fact that it is much easier to do things correctly from the beginning than to fix mistakes later. Another important point is a hymn to coordinated efforts. The first author was fully in charge of this project: after designing it, he has done all programming and managed a small team of monitors (which included training, control, and selection of the best ones to continue as the project progressed to more complicated stages), created a new structured database for his own future statistical analyses and research, wrote an official report (97 pages), and actually presented this project to the FDA during a one day meeting. As a result, the database that could not be used for years was restored, accepted, and served as the basis for future research. The number of common mistakes that were encountered, analyzed and corrected during this project is enormous; many of them were made because the teams responsible for the original conduct did not communicate well. For the first author it was

FIGURE 3.1
An initial position of a person interfacing the task of "cleaning of the data that cannot be cleaned"

an "eye opener", especially on communication and coordination. The main lesson learned can be formulated as follows: In order to successfully design and then finish a project of such complexity, the project lead should have a very deep understanding of *all* aspects of the project. Such goals cannot be achieved by bringing together even the best subject matter experts in separate fields if they don't understand what is going on in other fields.

In some sense, the situation discussed here highlights a real problem (especially in the big pharma model): after successful separation of duties in big projects (which, in fact, is a legitimate consequence of the technological progress), the task of intellectual coordination is often replaced by pure administrative coordination. Thus, even if parts of the process are performed perfectly, the final result can be disastrous. It would be fair to say that in 90% of the cases, the common approach (i.e. following established practice, rules and SOPs without questioning whether it will be sufficient) is more or less working; in 9% of the cases, there are serious problems that have a chance to be resolved only after investing a sufficient amount of time or money (much more than would be required with the proper organization), and in 1% of the cases, the current system just physically cannot produce any meaningful result. Moneywise, for a big CRO this might be even advantageous: proper organization requires highly skilled coordinators, which in 90% of cases are not needed; the extra cost of the remaining 9% is usually transferred to the client; and the remaining 1% is just ignored—the client is left with nothing, but money is usually paid. Let us name our second case study "*Cleaning of data that could not be cleaned.*" Figure 3.1 illustrates the first author's initial position when he interfaced this task.

Background: The data were collected and cleaned up for a pivotal phase III study with almost 700 patients from 38 sites conducted in the US, EU and South Africa (SA). The data collection was done by CRO#1 and the data cleaning process took more time and money than was originally expected (which means that the database was not among the 90% of cases that could safely pass on "auto pilot"). At some point, the client (small pharma) decided to switch to another vendor (CRO#2). Despite numerous outstanding non-resolved queries, the entire database was locked. CRO#2, after reviewing the entire study, refused (wisely, one should say) to continue cleaning, but retained all other activities for data management, regulatory, medical writing and, to some extent, biostatistics.

In order to finish data cleaning, the sponsor hired a small specialized biostatistics CRO#3 (formerly a biostatistics department of another big CRO that went out of business). The team at CRO#3 was highly skilled; its top biostatisticians did a great job designing the data cleaning process and creation of the structural database, while preparing a pretty complicated integrated summary of safety (ISS) database for an upcoming (in about 1 year) BLA that included 21 previously conducted RCTs and the study in question. It now occurred to the client (small pharma) that they were in strong need of the biostatistics department in house (still no DM!). Please hold your breath, because this is a typical development in small pharma. The data can be pouring in from 10–15 big and small CROs during 10 years of clinical development and the company may not even have a DM department. Furthermore, it may not even have a single permanent biostatistician position (with multiple contract statisticians working in periods of 3–4 months each) for several years!

The newly created biostatistics department consisted of a director of biostatistics and a senior statistical programmer (which happened to be the first author). During the final year prior to the BLA, the director was busy working on study reports for the ISS and the pivotal study, while the senior statistical programmer, with the help of CRO#3 and two biostatisticians from CRO#2, was preparing the entire programming package for the BLA. Eventually, the package was submitted to the FDA and, after 1 year of review, it was rejected with a long list of questions to be addressed, a quarter of which were related to the quality and integrity of the collected data.

Within the next few months, the director resigned, and the senior programmer was promoted to a senior biostatistician and remained the only biostatistician (no programmers, no DM) to help answer these questions. In a year, a newly promoted associate director (AD) of biostatistics finally understood how to clean the data that stalled all communications with the FDA, and proposed the solution to senior management, who were contemplating to go back to CRO#1 (despite all previous negative experience and strong arguments from the AD). A compromise was made that new cleaning would be done for $3,000,000 in 6 months (instead of $6,000,000 in 8 months as requested by CRO#1), under close supervision of, and with the full power of the AD. Luckily, the budget was really tight and the AD received the director title, $200,000 for the hired help, and 4 months to finish the project.

Intermission: This entire story is provided to show (as the newly appointed director learned later) an absolutely typical chain of events in a small pharma company and possible career paths for a biostatistician there. One may want to stay put *for years* (even decades), learn *everything* (medical science specific to the product, medical writing, regulatory requirements, negotiation, DM, programming, project management) and move up toward a celebrated position of director of biostatistics/data management/statistical programming. Alternatively, one may decide to change jobs more frequently, learn a little less and move up to a level of director of biostatistics or director of statistical programming. If one prefers to be narrowly specialized, he/she may decide to move to big pharma.

A description of the "stubborn" database: About 150 lab assays per patient; the number of required measurements varying from 0 (!) to almost 100; 40–50 local labs with paper reports (some sites had separate labs; e.g., for chemistry and hematology); 3 big central labs with electronically captured data for every region (one of them turned out to be a combo of 3 independent labs); data divided into 3 categories—chemistry, hematology and urine, and 3 regions—US, EU and SA; an average number of records per patient in the planned SDB about 1,500 (patient, test, date, time); taking into account all intermediate datasets to be checked, the number of fields subject to check is around 150,000,000. The database was being cleaned for years; the rate of errors from any sequential random audit stayed at 10–15%. The FDA requested 100% verification for this data to be considered for any reasonable reporting.

Normal technical difficulties: Just the numbers provided above warrant serious considerations: for some assays, dozens of observations per patient were collected; e.g. getting into the database the lab norms from 50+ labs applied to correct demographics of each patient are technically challenging.

A root cause analysis revealed the following major issues that prevented this data from being cleaned:

1. Multiple sources with distinctively different data collection— electronic vs. paper reports that were entered in paper CRF and then the CRF was the subject of data entry.

2. At some sites electronic records were not entered in the CRF, while at other sites they were. The design of data collection did not address this issue.

3. Data for required sampling points for chemistry and hematology was collected as "a rule" twice (!!!): one sample went to the central lab and the other one went to a local lab. When there was only one lab available, only one sample was collected.

4. There were certain rules on which of the two results from the same time point should be used in the CRF and/or for analyses, but they were different for the hematology (preference was set to the local lab) and for the chemistry (preference was set to the central lab).

5. The study design revealed a serious problem: two distinct time-lines for data collection. One timeline was standard: baseline, day 1, day 2, ..., day 7, ..., discharge, 6-week follow-up. The other timeline was built around transfusions: "pre-" and "post-" each transfusion. Transfusions were performed on an "as needed" basis, which means there was no common time scale across the patients, and the number of studied transfusions varied from 1 to 20.

6. The results from a single time point could be used for multiple study data points; for instance, one record could legitimately be used as the day 2, the "post-" for transfusion 3, and the "pre-" for transfusion 4.

7. The rules for choosing where the result should be entered were not well-thought out and, obviously, there was insufficient training of monitors. For example, records may fit for a couple or more infusions if they were back-to-back; for some sites (see item 2) the data were entered in the CRF while for others it was done programmatically later, when creating the final database.

8. The program that combined data from different sources into a single database was written by CRO#3 in a so-called "consulting" style— minimal comments, with (intentional or unintentional) implication "call us if you need anything". In addition, it was written by a *team*, which means there was no single responsible individual.

It took the AD several weeks of non-stop checks with tracing different groups of records to recreate the actual flow of the data. One may think that the AD was just an inexperienced programmer, but this was not the case. The real complexities were: i) 3 big, differently formatted, electronic regional databases divided into 3 lab categories (hematology, chemistry, urine), with different units and hard-coded conversions, and not always compatible assays with the same names, but performed by different instruments; ii) lab norms with different age divisions which were not necessarily stored in the main data; iii) data entered from the CRF with 40–50 lab norms datasets that were in some cases hard-coded, and in others read from Excel; iv) application of the rules for assigning data points for the electronically collected data; selection of data for analyses; handling of duplicates when an electronic record meets itself if it was entered in the CRF, etc, etc.

9. The created by CRO#3 ERRATA was obtained without any respect to the data flow based solely on resolution of queries by monitors, who were working not with electronic data, but rather with the CRF and the source documents.

10. All of the above created the following (frustrating) situation. The monitor (unaware of any of the nine aforementioned problems) would find a discrepancy between reporting of some result and a source document with this result (that could be either a printout from a local or the central lab) and would fill out the data clarification form (DCF), where "small details" of the origin of the record, the presence of duplicates, etc. were not even mentioned. The programmer from the CRO (unaware of the lab generation programming package) would enter the proposed change without any edits into ERRATA, and then would apply this correction to the final database. If this record had a duplicate (very likely), nothing would happen at all, because the duplicate would replace the corrected record. Of course, if a change requires 3–5 iterations, we should call it quits.

Assessment of the situation: Continuing data cleaning was out of question. Addressing all resolved queries properly in the present setting was possible, but required an incredible amount of work; in addition, during previous years of continuous cleaning, the database was messed up completely. Thus, everything that was done in the recent few years, starting from the lab generation program and the ERRATA should go. The solution proposed by CRO#1 was stunning: "Let's do it again, starting from the data entry." Let us accentuate that this is the most typical and most stupid decision. They did not know what went wrong the first time; they just admitted that it went really wrong. They had no "lessons learned" and no idea of how to avoid a similar situation again. Nevertheless, "Let's do it again!" After several rounds of discussion with CRO#1, they agreed to accept some supervision from the client, but still proposed to start from repeating the data entry. (Sure, this is the thing that they could do well and make money along the way!)

Let us look at what actually happened (Figure 3.2). A problematic part was the transition $\boxed{\text{CRF}} \rightarrow \boxed{\text{CRF DB}}$. There were no indications that data entry was done badly; moreover, repeating the data entry would not fix anything. The real problem was the data recorded in the CRF. Thus, to fix the situation,

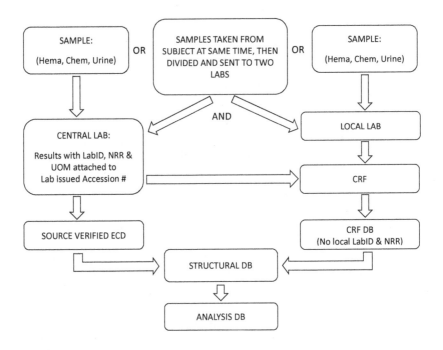

FIGURE 3.2
Actual data collection process

we would have to go very deep and re-do the CRF design and entry, with supplying at least the "LabID" as an additional field. This would, in fact, require re-opening of the completed study and all locked databases, and having an extremely skilled team of monitors to fill in the CRF on top of normal data entry. The only possible (significant) savings could come from not entering the ECD into the CRF. However, under such a scenario, we would face numerous compliance and regulatory issues, and we would end up with the CRF and the corresponding CRF DB "inferior" (i.e. less data) to the existing ones. Thus, despite a huge amount of required resources (personnel, money, and time), this would not guarantee the final desired outcome (restoration of credibility of the database).

A side note: Going back to Figure 3.2, it is very instructive to look at the flow of electronically captured data (ECD) in order to learn a lesson and avoid such mistakes at the design stage at any cost. The ECD was generated in the central lab, and then a (randomly) selected part of it was manually entered into the CRF (transition $\boxed{\text{CENTRAL LAB}} \rightarrow \boxed{\text{CRF}}$), with inevitable

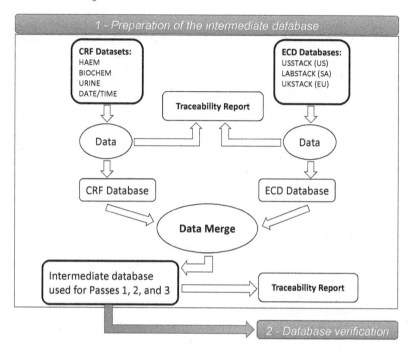

FIGURE 3.3
Preparation of the intermediate database

creation of duplicates and "loss of identity" due to the data from local labs which were manually entered into the electronic data base. Thus, the ECD meets itself (when combining SOURCE VERIFIED ECD and CRF DB to produce STRUCTURAL DB), with no means to even recognize it. Without exaggeration, such a "design" is prone to error-ridden execution.

The solution: The only solution that did not not require opening a can of worms (finished and closed study) was to start from the locked original database, still separated in the ECD and the CRF, and then combine them properly, achieving 100% source verification in the process. The most interesting question here is "What are the source documents?" The good news is that the ECD does not need any verification—*it is the source by itself.* For the data reported by local labs, the source documentation is the existing printouts of lab reports.

The only remaining problem is the mix of the ECD and local lab data without any identification (and multiple duplicates for both types). Thus, the first priority is to separate ECD and local lab data in the CRF DB. Then we can verify the ECD part against its original version (it will be needed after two data entries and intensive cleaning!) and verify the local lab part against printouts from the local labs. During verification of local labs we should be able to identify a lab that performed tests and enter LabID, which later permits us to derive the normal range reference (NRR). Then we can create the reporting database containing all collected data (STRUCTURAL DB) and then use it together with the rules specified in the study protocol to create the analysis database (ANALYSIS DB).

This sounds like a plan, but there is one small detail: how do we separate ECD records from the rest in the CRF DB? In the ideal world, we should be able to match parts of the CRF DB programmatically to the records in the ECD by common variables: PATNO, DATETIME, ASSAY, VALUE, UNIT, which should provide unique identification. Whatever is left should be viewed as the local lab data and could be verified by a relatively small team of trained monitors who should have the means to fix mistakes of data entry and enter lab identification. In the real world, we should be prepared for anything, at least mentally. Not all contingency plans are practical; sometimes it pays off to just be flexible and implement a solution only for the encountered problem—after all, no one can foresee everything!

Strictly speaking, the rest of the solution was a set of technical details.

Special attention was paid to *transparency* and *traceability*. Every confirmation and any change was properly documented. Every record from the starting place in the previously locked database was made traceable to the reporting database and vice versa. The reliable vehicle for entering fixes was developed; this helped to avoid buying expensive software and spending months on its validation. The file structure to store all project-related information was designed.

Another challenge was the maximum automation of the process. There was no other (reasonable) choice but to store all relevant data in the SAS format. For data entry, we settled on Excel. The principle was simple: a SAS data set that was subject of review was converted into an Excel spreadsheet where all fixes and entries were made into newly (programmatically) created fields; the spreadsheet was stored under a new name (e.g. letter "R" was added to indicate "Reviewed") and was read back into SAS (again with a new name). All edit checks that were designed to assure consistency and accuracy were run on SAS data using SAS programs. All individual datasets that returned discrepancies were reviewed from the moment when they were stored until all discrepancies were resolved.

The entire process was designed with multiple steps. SOPs and training manuals were created. All these materials with multiple memos, intermediate reports, rationales, etc. were stored in the Word format into a specially designed structure. When the project was finished, all supportive documentation (such as SAS datasets, Word documents and Excel workbooks which were used to report non-repeating activities) were locked (made "Read only") after they were finalized.

In total, there were more than 30,000 SAS datasets and Excel tables, located in almost 4,000 folders, generated during this project. Locking such

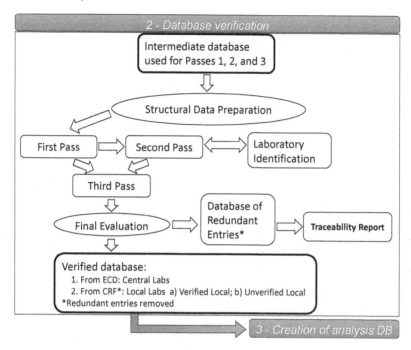

FIGURE 3.4
Database verification

an amount of data without automation was highly undesirable; therefore a special program DATALOCK was written to perform this task.

Of course this is a very general description of the entire project. There were hundreds of small (and not so small) issues, problems and challenges along the way (awaited, understood and resolved). In order to get a glimpse into a general project flow, let us skim through the entire process.

Figure 3.3 is a schematic of the first stage (preparation of the intermediate database). As we can see from this figure, both CRF and ECD databases had to be constructed first. Clearly, CRF reporting date/time could have been problematic if two samples were drawn at different times. Traceability reports had to deal with different dataset formats, removal of data from patients who did not participate in the study, etc.

Figure 3.4 shows a schematic of the second stage (database verification). Here, the "First Pass" was planned to verify possible matches between ECD records of the CRF database and the ECD database. This was necessary since a fuzzy match could occur due to the fact that the ECD records underwent data entry and cleaning twice. The "Second Pass" was planned to verify remaining records against the printouts. This was more than a usual cleaning—it was 100% verification, with correction of all discrepancies created during data entry and data re-structuring. Also, originally it was planned to have the final review (e.g. to clean any remaining mess by resolving the cases that were too hard for the team of monitors), by one or two subject matter experts familiar with the entire process. In reality, the remaining mess after two passes was either too big or too simple for the subject matter experts, and therefore a selected small team of monitors was trained to perform the "Third Pass" (as it turned out, they were able to resolve > 90% of the outstanding issues), and only then was the "Final Review" performed.

Figure 3.5 shows a schematic of the third stage (creation of the analysis database). During this stage, the first goal was to deliver the normal range reference for every result in the CRF originated from a local lab, and convert the results into standard units. For some tests, the unit field was filled before data entry (actually printed). As one could expect, at the sites where the units were different from the pre-specified ones, some results were converted while others were not (depending on the qualification of a specific monitor). This work originated in the second stage (cf. Figure 3.4), but, in fact, it could be completed only after all information on local labs was pulled together. In some cases, before making a conversion of the results, some fixes had to be applied to the "Final Review" from the second stage, which again confirms the iterative nature of any DM project.

Most of the difficulties during the third stage were absolutely standard for laboratory databases created for international multi-center trials: a variety of normal range references, different tests with the same names, the same tests with different names, incompatible tests due to different instruments, a variety of units for a given test, completely missing documentation for some sites and/or labs, different formatting for different central labs, different conventions for reporting (e.g. some labs would put into their printouts date/time of processing instead of date/time of sample collection), and reporting of missing times (e.g. at some labs it would be set to missing, whereas at others it would be set to 00.00.00).

Figure 3.6 shows the organization of the final created and stored (locked) data from 688 patients from 46 sites. The "housekeeping" was quite challenging as well. Each sub-folder inside the "Subject" folder contained a pre-defined structure for the three stages of data processing, separately for 3 lab groups. On top of permanently stored: SAS_DS_in, Excel_Sheet_in, Excel_Sheet_out, SAS_DS_out, there were plenty of intermediate reports of edit checks, a number of unresolved issues, etc. In addition, there was a temporary folder to keep the summary of edit checks throughout the entire database. All of these datasets were removed when they became empty (after every discrepancy was resolved).

Finally, it should be noted that programming of this project required some serious digging into SAS Component Language (SCL) and direct access to Windows using the X command, but the time and efforts spent were well paid off.

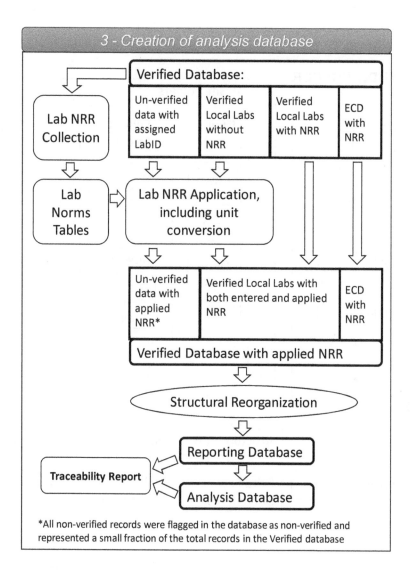

FIGURE 3.5
Creation of the analysis database

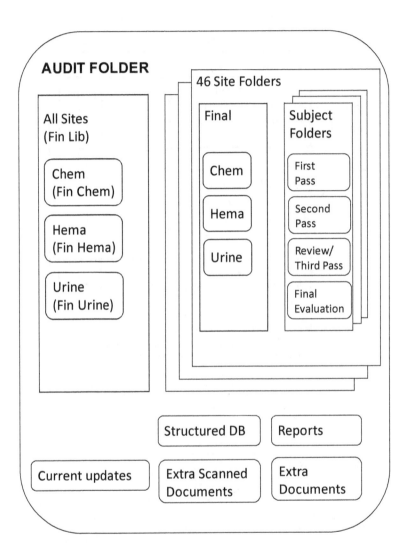

FIGURE 3.6
The structure of created and stored (locked) data

3.8 Summary

The purpose of this chapter was to demonstrate (especially through the case study in §3.7) that everything is possible, but in order to complete a data project of high complexity, it is crucial to have real multi-field expertise (not necessarily in one person): exceptional programming skills, deep understanding of data management principles and regulatory requirements, very good knowledge of laboratory data (specific tests) and practice, decent management/organizational skills. What has not been mentioned?—Biostatistics. But the biostatistics actually dictates what is needed—the final reliable analytical database which, in fact, shapes the requirements for the planned outcome of such projects.

In order to design and coordinate such a project, it is necessary to have a "data hero". Please, forgive blasphemy, but this is not a team job! Theoretically it is possible to have in one place a designated team of subject matter experts for a long period of time, let them interchange with each other all relevant experience in terms of "what my field can do for the project" and wait for a synergetic summary result (because a simple sum of possibilities would not permit a satisfactory design). In reality, this is very unlikely to happen unless a "data hero" is present.

What are the requirements for this "data hero"? A good (not necessary exceptional) knowledge based on understanding of main principles in all neighboring fields, a broad experience in solving problems rooted in several fields simultaneously, and an ability to dig out a specific solution or expertise for any encountered challenge inside each field.

Who is best positioned to be such a "data hero"?—Yes, the biostatistician. And, yes, his/her title could also be the statistical programmer. Has one ever seen a "pure" data manager who can do it?—We have not.

4

Biostatistics

4.1 Introduction

As was mentioned previously, there are some major differences in the work of a biostatistician in small and big pharma companies. The difference stems from the nature of the two business models. A small biotechnology company may just focus on development of a single compound. This may be the company's first major product, and the "life and death" of the company may depend on whether the compound succeeds or not. The biostatistician (if present) may be charged with doing all data tasks, including data management, statistical programming, design and analysis of clinical trials, etc.

By contrast, the big pharma model operates on a very different scale. The number of ongoing projects/studies is incomparably larger; multiple submissions as well as in-licensing/out-licensing of compounds may occur simultaneously; therefore, many more people are employed to run the business. An advantage of the big pharma model is that the tasks are split among various functions in such a way that more work can be accomplished, and potentially more successes can be obtained. Also, big pharma companies are generally more robust to development failures than small pharma. A disadvantage of the big pharma is rooted in its scale—due to complexity in the organization, big companies may be less flexible and may require more time to respond to emerging business challenges compared to small companies.

To appreciate the massive amount of work and complexity of the big pharma model, it is instructive to look at it through a "magnifying glass".

Consider a hypothetical big pharma company which performs very stably financially and has a well-diversified and balanced portfolio of drug candidate compounds. Suppose the company's drug development pipeline is categorized into the following ten therapeutic areas:

- Oncology

- Immunology

- Cardiovascular

- Metabolic disorders

- Ophthalmology

- Neurology

- Musculoskeletal disorders

- Virology

- Infectious diseases

- Rare diseases

Within each therapeutic area, there can be several diseases of special interest. For instance, in neurology the list of diseases may include: 1) neurodegenerative disorders, such as Alzheimer's disease, Parkinson's disease, progressive supranuclear palsy, etc.; 2) neuroinflammatory disorders, such as multiple sclerosis; 3) other unmet medical needs such as major depressive disorder, schizophrenia; etc.

Let us now consider a particular neurological disorder, e.g. Alzheimer's disease. In the company pipeline there could be several drug candidate compounds which have different mechanisms of action to treat the disease; e.g. gamma-secretase inhibitors, BACE inhibitors, microtubule stabilizers, etc. Also, several disease indications may have similar pathogenesis, and the same drug candidate may be tested across different indications.

For a given compound, there is usually a clinical development plan, a carefully crafted strategic document describing a sequence of clinical trials, each with its own specific objectives that would hopefully generate sufficient evidence that the investigational drug is safe and efficacious for treatment/management of the disease. As shown in Figure 1.4, the clinical development process is commonly split into three phases (I, II, III). If the compound successfully passes all three phases, a regulatory dossier summarizing all data from the conducted studies is submitted to the health authorities in request of marketing authorization. If the drug is approved and the marketing authorization is granted, the patients will soon start benefiting from the new therapy. However, even after the approval, there will be further follow-up (post-marketing) studies to investigate long-term safety and adverse events of the drug.

The number and types of clinical trials within a clinical development program may vary depending on the disease and/or the compound under investigation. This number can count tens (and up to a hundred) studies.

Phase I usually starts with first-in-human dose escalation studies to assess safety, tolerability, pharmacokinetics (PK), pharmacodynamics (PD), as well as early signals of efficacy (target engagement) of the compound. One of the main goals of phase I is to establish the maximum tolerated dose (MTD), which is the upper bound on the range of doses for subsequent studies. In addition, phase I may include various clinical PK studies [83], such as a study to characterize absorption, distribution, metabolism and excretion (ADME), bioavailability/bioequivalence studies, drug-drug interaction (DDI) studies, and possibly PK studies in special populations such as patients with renal/hepatic impairment, elderly, pediatric, and ethnic subgroups. The clinical pharmacology studies (although classified as "phase I") may be conducted in later development phases, as deemed appropriate. The total number of PK studies within a given clinical development program can be around 20–30.

Phase II is arguably one of the most important and challenging parts in the entire development process [97], because it bridges the exploratory (phase I) and the confirmatory (phase III) stages. Ideally, at the end of phase II we want to have a high probability of making the correct decision on whether to continue the development of the compound (by investing in large and expensive phase III pivotal trials), or stop the development. Some important research questions to be addressed in phase II are as follows [90]:

- Is there any evidence of the drug effect ("proof-of-concept")?

- Which dose(s) exhibit(s) a response different from the control?

- What is the dose–response relationship?

- What is the "optimal" dose?

Some statistical methodologies to address the above questions have been discussed in the literature [16]. A common clinical development program may consist of up to 5 (maybe more) phase II studies, including PoC, dose-ranging, and dose/regimen calibration trials.

Finally, in phase III, the goal is to make a formal comparison of efficacy of the new experimental therapy vs. standard-of-care (or placebo) by testing a pre-specified clinical research hypothesis. The health authorities typically require evidence from two independent phase III pivotal studies demonstrating consistent clinical efficacy results. The drug should also exhibit a favorable risk/benefit profile, and in some cases, it is also mandatory to demonstrate cost-efficiency [76].

Note: Here we discuss only clinical development, without delving into drug discovery and pre-clinical development. Undoubtedly the in-vitro and in-vivo experiments performed in very early stages (pre-clinically) require a lot of time, effort, scientific knowledge and expertise and call for high-quality statistical support as well. They merit a separate discussion which is beyond the scope of the current chapter. A good recent reference for nonclinical statistical support is [107].

All of the aforementioned indicates that in the big pharma model there is a strong need for constant careful review, calibration, and prioritization of the development portfolio (within disease areas and across the organization). Most promising projects should be prioritized and taken forward; unsuccessful compounds should be terminated; and "reasonable" projects should be ongoing at a slower pace or "parked" for better times.

On top of the internal project-related work, there may be some additional strategic activities that our hypothetical big pharma company is pursuing. These include, but are not limited to:

- Partnerships with other companies via in-licensing/out-licensing of compounds (e.g. in-licensing would require a careful due diligence to ensure that the compound is indeed promising).

- Acquisition of small companies that have achieved some level of success in the development of compounds for high unmet medical needs.

- Combining development efforts with other global pharmaceutical companies by performing platform/umbrella/basket trials [86].

- Diversifying business by partnering with IT companies to develop medical diagnostic tools and digital therapeutics.

- Looking into "real world evidence" (RWE) as an alternative tool to randomized clinical trials.

Finally, any drug development enterprise (regardless of whether it is conducted by small or big pharma) must comply with regulatory standards to ensure high quality, transparency and credibility.

In summary, some important highlights about big pharma can be made as follows:

- In the big pharma model, the number of projects/studies and the amount of corresponding work can be astronomical.

- Given the scope of work, doing everything in house may be infeasible. This calls for some strategic outsourcing of certain activities to CROs.

- Staying in business and surviving the competition requires a carefully planned and consistent business model. The close coordination of efforts, continuous monitoring and prioritization of work are required across the entire company and all collaborating CROs.

- *Communication* among various functions (e.g., scientific, regulatory, operational, business planning and execution, just to name a few) must be agile, efficient, and transparent.

- *Compliance* with legal and regulatory standards must be flawless; otherwise there can be a risk of major problems, including significant financial and other penalties to the company.

Perhaps no one in the world would be able to provide a universal solution on how to optimize the drug development process for a given company or across the entire industry. For this to be possible, a "universal problem solver" would have to have certain knowledge of the past, the present, and the future.

There are some very good books and scientific publications on pharmaceutical research planning and portfolio optimization; see Bergman and Gittins [10], Burman and Senn [18] and references therein.

In the current chapter we do not intend to have a discussion at such a global level. After all, the focus of this book is on biostatistics and the pragmatic approach. In what follows, we shall elaborate on the role of a biostatistician in clinical drug development and the type of impact that can be achieved.

4.2 The biostatistician's role

To start, let us make an important statement that the role of the biostatistician in the drug development enterprise is not restricted to the clinical phases. Or as Grieve [52] puts it: *"Pharmaceutical statistics is not just clinical statistics."* Given the breadth of the spectrum of pharmaceutical activities (cf. Figures 1.3 and 1.4), there is a need for statistical support in practically all stages, starting from drug discovery and up to commercialization, marketing, and post-approval development.

Below is a snapshot (not all-inclusive) of the areas that require the engagement of biostatisticians:

1. Drug discovery/pre-clinical:

    – Genetics/genomics/proteomics

    – Toxicology

    – Process improvement

    – Quality control

    – Determination of shelf-life

2. Clinical:

    – Phase I–III studies

    – PK/PD modeling and simulation

    – Regulatory support

    – Publication support

    – Portfolio management

3. Regulatory submissions:

    – Integrated summaries of safety and efficacy (ISS/ISE)

    – Development of the common technical document (CTD)

4. Post-approval:

- Marketing support
- Post-marketing surveillance
- Outcomes research
- Epidemiology

Note that some activities may be applicable across different stages; e.g. PK/PD modeling and simulation are applied to both animal (pre-clinical) and human (clinical) data; publication support is required regardless of whether data comes from pre-clinical, early/late clinical, or post-marketing research; regulatory statistics are relevant in toxicology, pharmaceutical and clinical development, etc.

High-quality statistical support can be especially valuable when a project reaches strategic transition points, such as transitions pre-clinical $\rightarrow$ clinical, phase I $\rightarrow$ phase II, and phase II $\rightarrow$ phase III (Figure 1.4). Quantifying go/no-go decision criteria and using pre-specified decision rules with known statistical properties can introduce rigor and improve quality of the decision-making process.

Based on what has just been described, a biostatistician in a big pharma company faces an important strategic question: "What area of drug development do I want to work in?" The answer will depend on personal preference, actual open positions, level of experience, etc. It is a historical quirk that most statisticians in the pharmaceutical industry work largely in the clinical phases of drug development [53]. However, clinical development itself is very broad and requires choices to be made.

The therapeutic area is perhaps the first important consideration. For instance, clinical trials in oncology have a lot of special features (e.g. trial designs, endpoints, statistical methodologies for data analysis) that make the oncology statistician a distinct profession.

Another important consideration is the choice between early clinical development and late clinical (full) development. Early clinical development is exploratory in nature; the clinical trials aim at assessing safety, tolerability, PK, PD, and generating new research hypotheses to be tested in subsequent pivotal studies; the trials are relatively small ($\sim$20–100 subjects) and run fast; and the number of trials within a given program can be quite large. By contrast, the full development is confirmatory in nature; clinical trials are large (several hundreds or thousands of patients) and long in duration. Therefore, an early development statistician would typically support many small-scale studies, whereas a full development statistician would work on one or two large-scale studies.

It should be noted that both early and full clinical development are subject to regulation, and regulatory requirements for the quality are equally (and increasingly) high. After all, if a drug is successful, all conducted trials (regardless whether early or full development) will be part of a submissions package (NDA/BLA in the US). On the other hand, it is not uncommon that many trials/projects fail due to lack of safety, efficacy, etc. In this case the

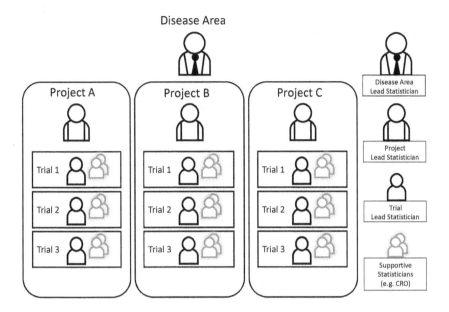

FIGURE 4.1

An example of the biostatistics operating model

clinical trial results would not carry the same level of importance as that of the successful projects. However, even for failed studies there are certain regulatory requirements for the results to be publicly disclosed at clinical trial registries, such as ClinicalTrials.gov in the US and/or European Clinical Trials Database (EudraCT) in Europe.

A distinct area that deserves special mention is statistical consulting. Some big pharma companies/CROs have designated statistical methodology groups that are tasked with solving complex problems across the entire spectrum of drug development, and advising clinical project teams on complex statistical issues. The statistical methodologists/consultants are expected to have excellent knowledge of state-of-the-art statistical techniques for modern drug development and hands-on experience applying them in practice. They are also frequently involved in development of new statistical methodologies, publishing research papers, providing various trainings to (non)-statisticians, etc. Statistical consultancy could be an excellent opportunity for a pharmaceutical statistician interested in problem solving in the broad sense. A major limitation is that such groups are not very common and the number of positions is usually quite limited.

Now, let us contextualize the biostatistician's role in a big pharma company by considering a particular example of an operating model. Note that this is only an example, which may or may not be optimal in different circumstances. We present it for illustration purposes only, to highlight the ideas and the logic behind why multiple statisticians are needed to support portfolios of projects.

A side note: Here we just consider project-related biostatistics support (which typically accounts for 80% or more of the job of individual contributors). We do not cover project management, participation in strategic initiatives, trainings, and other important job aspects. Also, we exclude activities of senior biostatistics leadership roles (e.g. at the level of VP and above)—these are highly strategic and are yet beyond the second author's experience.

To fix ideas, let us consider early clinical development (from the first-in-human (phase I) to the proof-of-concept (phase IIa)), and a particular disease area (one of the ten discussed in §4.1). Within the disease area we have several projects, and each project may consist of one or more studies.

Figure 4.1 shows a structure of the biostatistics operating model. It is common to have a designated *disease area lead statistician*. This individual (whose title is typically an associate director of biostatistics or above) would play a key role providing statistical support for most strategic activities in (early) clinical development within the given disease area. In addition to the technical expertise, some important responsibilities for this role include, but are not limited to:

- In close collaboration with major stakeholders (clinical project teams and management) develops and drives strategic quantitative frameworks for clinical development programs within the disease area.

- Partners with senior clinical leadership to negotiate and influence clinical development plans.

- Leads and facilitates teams in the definition and quantitative evaluation of competing program/trial/analysis.

- Provides technical statistical expertise and leadership to apply cutting edge statistical methodology, innovative designs, and creative project plan solutions within the assigned disease area.

- Provides biostatistics input into various decision boards in supporting quantitative decision making for reaching the target milestones.

- Represents biostatistics in all key interactions on the projects that take place with the health authorities.

The disease area lead statistician would be the main point of contact for any strategic biostatistics-related questions; he/she would have to maintain up-to-date knowledge of the planned and ongoing activities in the portfolio of projects within the assigned area, and would also coordinate the work of project-level and trial-level statisticians by making right prioritization of the ongoing activities. Also, depending on the company workload, the disease area lead statistician would frequently act as a project and/or trial statistician and interface various clinical project teams and senior clinical leaders.

For each project within the disease area, one biostatistician is designated to be the *project lead statistician*. The "project" may include all indications for a compound, or it may refer to a subset of selected indications. The project lead statistician would have to be a company employee who represents biostatistics on the research project team and various sub-teams; he/she would carry overall responsibility for statistical deliverables for the project, and would be closely involved in key studies (e.g. as the trial statistician). The project-level activities can be numerous and some of them may require a lot of time and effort. Below is just a snapshot of some important ones:

- Statistical input into clinical development plans and target product profiles.

- Statistical input into presentations for internal advisory/management boards.

- Drug safety update reports (DSURs)/Periodic safety update reports (PSURs).

- Statistical support for interactions with health authorities (input to briefing books and answers to questions).

- Pooling work (integrated summary of safety/efficacy).

- Project-level exploratory/ad-hoc data analyses.

- Project-level publications.

Finally, let us consider the trial-level biostatistics work, which is one of the main engines of any drug development program. For each trial, one statistician is assigned to be the *trial lead statistician*. This individual would have to be a company employee, ideally assigned to the study for the whole duration of the trial. Note that in practice many trials take years to complete, and statisticians often change jobs (within and outside the company). It is not uncommon that the design of the study, the development of a statistical analysis plan and related documents, the analysis of the data, and the writing of statistical sections of the clinical study report (CSR) are done by four different individuals who wore the same "trial lead statistician" hat at different time instants.

The trial lead statistician would be the statistical representative on the clinical trial team, would carry overall responsibility for statistical inputs and deliverables for the trial, would approve all trial-related documents that require statistical approval, and would be the statistical point of contact for most activities within the trial.

The trial-level biostatistics activities span all steps of the trial lifecycle including design, conduct, data analysis, reporting and dissemination of results. More specifically, throughout the course of the study, the trial lead statistician is involved in the following activities:

- Development of the trial protocol.

- Development of the treatment randomization schedule.

- Interfacing IRBs and health authorities.

- Input to the data monitoring committee (DMC) charter.

- Development of CRF.

- Development of statistical analysis plan, TLF shells, and first interpretable results (FIR) mock slide deck.

- Generation of interim and final data analysis results (and FIR).

- Writing statistical sections of the CSR.

- Statistical input to the clinical trial registry documents.

- Publication of trial results in clinical journal(s).

A distinct feature of our described operating model is the presence of *supportive statisticians* (cf. Figure 4.1) who may help the lead trial statisticians at different times on different parts of the trial. The supportive statisticians may be either internal or external (e.g. contractors or statisticians from CROs), and they may provide statistical support for selected, well-defined activities within the trial. These individuals are employed for good reasons which we explain momentarily.

Every trial-level activity starting from protocol development and ending with publication of the results is a multi-step approach. These steps can be broadly classified into 5 major categories: 1) main statistical contribution; 2) implementation (e.g. programming, documentation); 3) review cycle; 4) approval; and 5) interface with the study team. For steps 1), 3), 4), and 5), the natural responsibility in held by the trial lead statistician. While some of these steps are often tedious and time consuming, they are mandatory to meet the company SOPs and regulatory standards. Taking into account that a trial biostatistician usually supports multiple studies simultaneously, the total amount of work to be done can be very huge. As such, step 2) is frequently delegated to supportive statisticians. One commonly seen example of this paradigm is when the main trial documents (study protocol, statistical analysis plan, and related documents) are developed by a company trial lead statistician in house, the implementation (statistical input to CRF, programming of the TFLs, etc.) is outsourced to a CRO, the review and approval of the deliverables is performed in house, and the interface with the clinical study team is done by the company trial lead statistician.

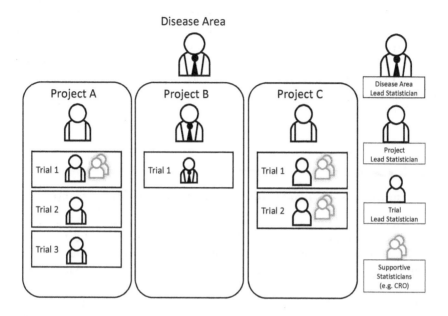

FIGURE 4.2
An example of a more realistic biostatistics operating model

If properly implemented, this model can potentially increase the company's capacity while maintaining the strict regulatory requirements for quality. To make it work (from the biostatistics perspective), it is essential to ensure clear understanding of the roles and responsibilities, and efficient implementation at all levels, including disease area, project, and trial. For the success of a given project, perhaps the most important part is communication between the project lead statistician and trial lead statistician(s). To ensure exchange of information, regular meetings and discussions are required, and the project lead statistician should be responsible for initiation of such meetings/discussions. The project lead statistician is also expected to be in contact with other line functions (as appropriate), for instance, project management, regulatory, clinical PK, etc.

At this point, a careful reader may wonder whether the described operating model is realistic. Indeed, different clinical development programs are usually at different development stages (some have a greater number of completed/planned trials than others), certain clinical projects carry greater importance than others, and various clinical project teams may compete (even within the same company). Therefore, a well-balanced schematic in Figure 4.1 may be oversimplification.

Figure 4.2 provides an example of a more realistic version of Figure 4.1, where the number of trials and the nature of biostatistical support vary across the projects. For Project A, the project lead statistician also "wears the hat" of the trial lead statistician for each of the three trials within the project, and only for trial 1 is the supportive statistician available. For Project B, there is only one trial, and the disease area lead statistician acts as both the project lead and the trial lead statistician, without any supportive statisticians. Finally, for Project C, there are two trials, each with a designated trial lead statistician and supportive statisticians, and these are overseen by a separate project lead statistician.

4.3   Background assessment: what do we start with?

For anyone who is about to start a new job as a biostatistician (especially if this is the first job after graduate school), it may be very helpful to reflect on their accumulated knowledge/skills that can be relevant. Undoubtedly, the foundation is good knowledge of statistics and good programming skills. While the drug development enterprise (especially in big pharma) is a team effort, there is high expectation that each individual contributor can perform the assigned tasks independently, in a timely manner, and efficiently. The ability to learn quickly, efficient communication, self-organization, time management, and other "soft" skills are certainly important ingredients; however, without due technical and scientific background knowledge (and professional expertise), any individual biostatistician will likely face big challenges delivering the required output.

For instance, the ICH E9 guidance on "Statistical Principles for Clinical Trials" [61] has a statement that *"...the actual responsibility for all statistical work associated with clinical trials will lie with an appropriately qualified and experienced statistician..."*, who *"...should have a combination of education/training and experience sufficient to implement the principles articulated in this guidance."* Such an explicit statement expresses the consensus that the statistics discipline is central to ensure reliable clinical trial results, and this is one of the reasons that statistical scientists continue to be in high demand in the pharmaceutical industry [52].

Several papers in the literature provide excellent elaborations on what it means to be a "qualified and experienced statistician" for clinical trials [22, 48, 104], and how one can develop the necessary knowledge and experience [62, 71, 98]. The good news is that a person with a background in mathematics and/or computer science (whose coursework also included statistics) already has a lot of knowledge and skills that are transferrable and could qualify them for a clinical trial statistician role.

In this section, we give a personal account of the statistics coursework that can be very helpful in the pharma industry. Clearly, this is a subjective presentation and should be interpreted as such. It is primarily based on the second author's experience obtaining an undergraduate degree in applied mathematics from an Eastern European country (during 1996–2001), and then completing a Masters program in statistics from a US university during 2002–2004. One can imagine that some of the ideas may already be outdated (due to rapid technological advances and increasing understanding that many problems can now be solved in seconds using the power of computers rather than analytically); however many principles still apply and a careful reader will hopefully find them useful.

The second author considers himself very lucky to have benefited from two distinct (if not to say, disjoint) systems of mathematical education: undergraduate Eastern European and graduate (Masters and then PhD) North American (US). The outcome from such a fusion has (unexpectedly) turned out to be truly synergistic.

The undergraduate 5-year program was in applied mathematics; the coursework was very intense, theoretical, and often intimidating. I still remember one professor of mathematical analysis who would say: "If I give you a problem and you solve it—this doesn't tell me anything; however, if you do NOT solve it, this tells me a lot..." At the end of a semester, the examinations were oral (face-to-face with the lecturer), closed book and closed notes. The questions could arise not only from the taught course, but also from adjacent areas, which means that in order to obtain the highest grade, an examinee would have to demonstrate very deep understanding of the subject and its connection to other fields. The system was designed to develop logical thinking, breadth of mathematical knowledge, and, to a lesser extent, real-life applications. The coursework over the 5 years included: mathematical analysis, linear and abstract algebra, analytic and differential geometry, ordinary and partial differential equations, functional analysis, discrete mathematics, optimization, numerical analysis, generalized functions, special functions, Sobolev spaces, operator theory, diffraction theory, mechanics, mathematical physics, probability, mathematical statistics, random processes, scientific programming, and game theory. At the end of the program, a student would also have to defend a thesis (similar to a M.Sc. in the US). After completion of the program (with such an impressive list of courses and acquired theoretical knowledge and almost zero practical experience), a natural question arose: "Where do I use all this?" A pragmatic decision was to pursue graduate studies in statistics.

The Masters program in statistics was a 2-year program in one of the US universities. For the second author, the major cultural difference (and initially challenge) was the approach to education. In the US, graduate programs are very well structured and have clearly defined prerequisites and learning objectives for each course. Big emphasis was made on active learning and independent work—systematic study/revision of the material taught in class, timely completion of homework and projects, and solving various problems in preparation for midterm and final exams. All examinations were written, and one's grade would be determined based on what one could solve in the allotted time. With proper effort, self-discipline, and practice, the chances to succeed were very high, and in 2 years, the target milestone (M.Sc. in statistics) was reached.

In the US system of education, the ability to solve problems quickly and efficiently is of great importance. Often times it is better if you solve a dozen of problems with a very brief explanation of the steps you have made than if you dig deeply and try to give a flawless elegant justification of all provided arguments. In a sense, this is very useful, because in real life, every business enterprise wants to have solutions for their problems, and the method you use is just your personal matter (as long as the solution makes sense). Quoting a famous American statistician, John Tukey (1915–2000): *"An approximate answer to the right question is worth a great deal more than a precise answer to the wrong question."* Developing skills to solve problems and get the correct results fast, learning how to perform well under the pressure of multiple tasks, and developing a strong sense of urgency can substantially improve one's thinking abilities. Graduate studies in statistics definitely provide a framework for developing such a mindset.

Let us now consider in detail a coursework in statistics (the list, of course, is not all-inclusive) that is likely be present in the arsenal of a person who completed a Master's degree in statistics. The knowledge extracted from these courses provides a time-tested valuable asset that can be capitalized upon in the pharma industry. The list is as follows:

Foundations of statistics:

1. Probability theory
2. Statistical inference
3. Applied statistics I (Linear regression models and generalized linear models)
4. Applied statistics II (Experimental design and survey sampling)

Additional statistical courses:

5. Multivariate statistics
6. Bayesian statistics
7. Nonparametric statistics
8. Computational statistics and simulation
9. Categorical data analysis
10. Survival analysis
11. Clinical trials
12. Bioassay

The first four courses form the foundation of many graduate programs in statistics around the world. In fact, comprehensive examinations to test whether a candidate is prepared to pursue a PhD degree in statistics is usually cast in two parts: the first one (theoretical statistics) covers probability theory and statistical inference, and the second one (applied statistics) covers regression techniques and design of experiments. The qualifying threshold is typically set at 80% for the PhD level pass and 60% for the Masters level pass.

Different universities/departments developed their own curricula for the first four foundational statistics courses (at different levels of mathematical complexity). For a useful unifying example, one can refer to the Graduate Diploma examinations provided by the Royal Statistical Society (RSS)—https://www.rss.org.uk/—which, unfortunately, were discontinued after 2017. For an applied statistician working in the pharma industry, looking at the past RSS examination papers and solutions can be a very rewarding experience. The papers give many problems from real-life applications, and illustrate how statistical techniques can be used to solve these problems. Furthermore, the RSS website provides valuable reading lists for each statistical discipline that was in the scope of the examinations.

The second set (additional statistics courses) provide an expansion of the first set, with more specialized statistical methodologies that can be utilized in various real-life applications.

Multivariate statistics provides methods for analyzing multivariate data. These methods are increasingly useful today, because in practice many different variables are collected on the same subject/unit and the resulting datasets are naturally multi-dimensional. Principal component analysis (PCA), linear discriminant analysis (LDA), factor analysis, classification problems, and many other data science and machine learning fashionable techniques have roots in multivariate statistics.

Bayesian statistics provides an elegant alternative approach to the classical (frequentist) inference. The classical approach treats model parameters as fixed unknown quantities that can be estimated based on data. In the Bayesian setting, model parameters are assumed to be random, with some prior distribution. Bayes' formula is used to combine prior with observed data to obtain the posterior distribution on which statistical inference is performed. In the pharmaceutical industry, Bayesian methods have found broad and useful applications [4, 54, 69], and the trend continues! One quick example is the use of historical data (from previous trials) to elicit an informative prior on the effect of placebo/standard of care, which can potentially lead to reduction in the requisite sample size for a new experiment.

Nonparametric statistics gives an extremely useful set of techniques for distribution-free inference. There are many practical examples when it is infeasible to assume that data comes from a normal distribution, and therefore conventional methods (e.g. two-sample t-test) may fail. With a nonparametric approach, statistical inference is made on ranks of observations rather than the actual observed values. The results are robust to model misspecification and potential outliers in the data, are easy to interpret, and in many circumstances nonparametric tests have very high efficiency relative to parametric tests. One of the best graduate-level textbooks on nonparametric statistics is *Introduction to the Theory of Nonparametric Statistics* by Randles and Wolfe (1979 Wiley edition) [85], which unfortunately came out of print and is now of limited availability.

Computational statistics and simulation is perhaps one of the best investments of time and efforts in any graduate statistics coursework. As was mentioned in §1.3, two distinct tools for new knowledge acquisition have traditionally been formal logic (deduction) and empirical observation (experiment). Computational statistics presents yet another (very powerful) tool for discovery: use of the computer to simulate data under different hypothetical experimental scenarios, and quantify the characteristics of the simulated processes and phenomena. In drug development, clinical scenario evaluation is one of the cornerstones of clinical trial optimization [7, 29]. Simulation of a clinical trial (or even an entire clinical development program) under standard to worst-case scenarios can help assess the likelihood of success and inform decision makers about associated risks and opportunities. Clinical scenario evaluation is becoming an increasingly common and desirable skill in biostatistics job descriptions.

It is difficult to come up with a set of topics in computational statistics that deserve learning and a sequence in which they should be learned—the list will be very long and will require careful customization. Definitely, different methods for random number generation, bootstrap, numerical integration techniques, the Expectation-Maximization (EM) algorithm, Markov Chain Monte Carlo (MCMC) methods, the Metropolis-Hastings algorithm, the Gibbs sampler, and many more topics are all indispensable in a statistical scientist's arsenal. Knowledge of data mining techniques—CART, MARS, TreeNet, Random Forest—is a big plus. Knowledge of statistical software packages for computational statistics is in good analogy with a conventional wisdom "As many languages you know, as many times you are a human being." R, Python, MATLAB, BUGS—and the list is not exhaustive by any means. The second author's personal choice is the R language for many good reasons that will be mentioned shortly. As technology continues to rapidly evolve, computational power is getting cheaper and more affordable; thus the value of computational statistics will sustain.

Categorical data analysis is an excellent applied statistics course which provides techniques for analyzing categorical response variables in biostatistics and epidemiologic studies. If there are no explanatory variables (covariates), we have analysis of simple contingency tables. With covariates, the methods include the logistic regression model (for binary response), proportional odds model and continuation-ratio model (for ordinal categorical response), and multinomial logit model (for polychotomous nominal response). In addition, *log-linear* models can be very useful when there are two or more categorical response variables.

Survival analysis (or analysis of time-to-event data) is a crucial course for clinical trial statisticians, especially those who work in the area of clinical oncology. In many oncology trials, the primary outcome is either progression-free survival (time from randomization to disease progression or death from any cause) or overall survival (time from randomization to death from any cause). These times are likely to be censored in practice. How to properly analyze censored time-to-event data is an interesting and challenging question. Methods of survival analysis can be classified into *nonparametric* (e.g. Kaplan-Meier estimate; log-rank test), *parametric* (e.g. accelerated failure time model; proportional odds model), and *semiparametric* (Cox proportional hazards model). A good knowledge of these methods, together with hands-on experience of SAS PROC LIFETEST and PROC LIFEREG is a big asset of any biostatistician.

A course in *clinical trials* is taught at many biostatistics departments (and less commonly in statistics departments) in the US. This course, if available, simply cannot be missed. It equips a student with the knowledge of clinical research, types of clinical trial designs, randomization methods, approaches for data analysis, and much more (depending on the lecturer's expertise and time). This course can be taught at different levels of mathematical rigor: e.g. minimal math (just concepts and applications) or highly theoretical (including large-sample theory and martingales). The second author was extremely lucky to take this course from a professor who later became his PhD advisor and later a good friend. The book that formed the basis for motivation to become a professional biostatistician is *Randomization in Clinical Trials: Theory and Practice* by Rosenberger and Lachin [88]. An interested reader may check details, but it would not be an exaggeration to say that this course gave "tickets to life" to dozens (or even hundreds?) of graduate students who pursued their statistician careers in academia and industry.

Bioassay is a collection of methods for estimating dose–response relationships using generalized linear models, such as logistic and probit regression models. One of the central problems in bioassay is estimation of various quantiles of the dose–response curve, such as median effective dose (ED50), which is the dose for which probability of response is 50%. Applications of bioassay in the pharmaceutical industry are numerous: toxicology studies, relative potency analysis (to provide a comparison between two substances), animal carcinogenicity studies, etc. One important application of bioassay is in clinical oncology phase I trials to estimate the maximum tolerated dose (MTD). The MTD is often defined as a quantile of a monotone dose–toxicity probability curve, which is to be estimated sequentially, by using adaptive dose escalation designs. A proper application of bioassay methodologies can help an investigator to estimate the dose–toxicity relationship, the toxicity probabilities at the study doses, the MTD and other parameters of interest. This information is used for calibrating dose range for subsequent (phase II) studies.

It should be noted that the aforementioned courses may appear in different variations. In Europe, the RSS Graduate Diploma had a modular version of the examinations until 2009. This modular version included the Options Paper that contained six options: A) Statistics for Economics; B) Econometrics; C) Operational Research; D) Medical Statistics; E) Biometry; and F) Statistics for Industry and Quality Improvement.

The term *medical statistics* is widely used in the UK, whereas *biostatistics* is a more common term in the US. The medical statistics paper (option D) included the following topics: clinical trials, epidemiology, diagnostic tests, analysis of survival data, and health information. The biometry paper (option E) included topics in experimental design, survey methods, regression modeling, and bioassay. All these topics are highly relevant to pharmaceutical statisticians.

To conclude the discussion in this section, there are many other statistics courses that may become handy (sometimes in quite unexpected circumstances). For example, the second author regrets that he did not take the time series course in his Masters program. While time series analysis is certainly a powerful tool in econometrics and finance, it is not that very common in clinical trials. However, having this knowledge would help answer the following question: Why do we use so frequently the mixed-effects model for repeated measures (MMRM) and not the time series? It would never be too late to take a time series course, but this again exemplifies the old wisdom that "everything is good in its time" and that "knowledge is no burden".

4.4   A minimal sufficient set of tools for the biostatistician

In §4.3 we discussed some useful background knowledge one may possess just as an outcome of (graduate) studies in statistics. However, this academic knowledge may still be insufficient for full performance in the pharma industry. What is essential and what is not? How do we separate the wheat from the chaff? From the theory of statistical inference, the *sufficiency* principle posits that a *sufficient statistic* is one that, in a certain sense, captures all the information about the parameter of interest contained in the sample. A *minimal sufficient statistic* is one that achieves the most data reduction without loss of information about the parameter of interest. In this section, we make an attempt to elicit a "minimal sufficient" set of skills/knowledge that are required for a biostatistician in the pharma industry. This goal is in good alignment with the pragmatic approach philosophy.

Let us again refer to the ICH E9 guidance—an authoritative global reference (harmonized across different geographic regions, including the US, Europe, and Japan) for statistical principles for clinical trials [61]. This 38-page document provides a useful source not only for the statisticians in the pharma industry, but also for their scientific collaborators. Importantly, the guidance focuses on the *principles*, and not on any specific statistical procedures or methods. Just from its table of contents, it becomes clear that biostatistics is a pivotal element of any clinical development program, and biostatisticians are expected to play a major role in the design, conduct, analysis of data, and reporting of clinical trial results. The specific tasks to be accomplished during the process are numerous and will depend on many factors, including disease area, unmet medical need, competitive landscape, etc. To put some systematic framework around the search, we consider the following seven categories:

- Knowledge of the disease area (§4.4.1).

- Knowledge of the regulatory landscape (§4.4.2).

- Understanding of the clinical trial protocol (§4.4.3).

- Knowledge of statistical methodologies for protocol development (§4.4.4).

- Statistical software to enable protocol development/implementation (§4.4.5).

- Communication skills (to interact with the clinical trial team) (§4.4.6).

- Knowledge of processes (business operating model) (§4.4.7).

Let us now explore this list in more detail.

4.4.1   Knowledge of the disease area

The "executive summary" section of most project proposals that seek endorsement from upper management necessarily includes the description of a compound, its target mechanism of action, unmet medical need, existing animal models, translational and clinical approaches, etc. This background information is very crucial for making strategic decisions. A biostatistician may not have deep knowledge of all underlying science and biology, but, at the minimum, should understand main concepts (scientific, medical, and commercial rationales).

In translational medicine, the three key independent pillars have emerged as industry best practices [74]: the target, the dose, and the patient. These pillars are helpful benchmarks to keep in mind during work on a given project/trial. The *target* pillar corresponds to the biological aspect (do we have the right biological target for the selected disease?) The *dose* pillar corresponds to the pharmacological aspect (do we have the right dose/regimen that delivers the right exposure for the given biological target to exert the desired modulation over the stated time period without compromising patient safety?) Finally, the *patient* pillar corresponds to the target population (do we have the right subgroup of patients with the selected disease who would benefit from our drug?) To achieve a successful outcome, we should ensure that our clinical development programs are designed to accrue evidence (through science and careful experimentation) that we have a reliable biological target, a (nearly) optimal dose/regimen in the right patient population.

One useful exercise is to create a PowerPoint slide deck (just for your own use) with a description (in your own words) of the disease area, disease characteristics (biology), epidemiology, causes/risk factors, signs and symptoms, existing treatment(s), main strategies for new treatment development, competitive landscape (compounds that are in development by other companies, e.g. by researching ClinicalTrials.gov), and compounds that are currently in development in your company. Creating such a deck is by itself a rewarding intellectual exercise, and once it is created, ask your medical colleague to review it for consistency, to eliminate any major errors and to add some key elements that may have been missed—it will now become a highly valuable asset for your reference.

4.4.2   Knowledge of the regulatory landscape

It may now sound absurd, but the pharmaceutical industry was essentially unregulated until the early 20th century. Formation of the FDA in the US (with the passage of the 1906 Pure Food and Drug Act) gave a start to rigorous quality control and constantly increasing standards for prescription drugs. Today, the US FDA's mission is to protect the public health by "*...ensuring the safety, efficacy, and security of human and veterinary drugs, biological products, and medical devices; and by ensuring the safety of our nation's food supply, cosmetics, and products that emit radiation.*" In other countries throughout the world, the regulatory agencies are: the European Medicines Agency (EMA) in the EU, the Pharmaceutical and Medical Devices Agency (PMDA) in Japan, the Therapeutics Products Directorate (TPD) in Canada, etc. It would not be an exaggeration to say that today's pharmaceutical R&D is a highly regulated industry, with a diverse (country-specific) set of requirements for new drugs, biologics and medical devices.

A major step towards harmonization around the globe was made by formation of the International Conference on Harmonisation (ICH) in Brussels in 1990. The ICH brought together representatives of the health authorities and pharmaceutical industry in Europe, Japan and the US to discuss scientific and technical aspects of drug registration and achieve global alignment on the key requirements for safety, efficacy, and quality of drugs. The ICH issued many guidance documents which can be viewed as international standards for drug development. The ICH guidelines are categorized into four major types (https://www.ich.org/products/guidelines.html): Q for "quality" (manufacturing); S for "safety" (nonclinical); E for "efficacy" (clinical); and

M for "multidisciplinary". Our personal choice of the most relevant guidelines is:

- E9 – Statistical Principles for Clinical Trials.

- E8 – General Considerations for Clinical Trials.

- E10 – Choice of Control Group and Related Issues in Clinical Trials.

- E4 – Dose–Response Information to Support Drug Registration.

- E6 – Good Clinical Practice: Consolidated Guideline.

There are many useful country-specific regulatory documents. For instance, the FDA posts its guidance documents at

https://www.fda.gov/regulatoryinformation/guidances/default.Htm.

As of November 20th 2018, there are 2,679 of them; some are in the draft form (and open for comment to the public), and others are already in the final form. One useful important example is the guidance on adaptive designs for clinical trials of drugs and biologics [45]. Its first draft version was issued in 2010. Eight years later we have a substantially revised version (still in the draft form) which reflects the current thinking of the FDA on adaptive clinical trials.

In Europe, many useful guidelines for clinical biostatistics are available from EMA at https://www.ema.europa.eu/en/human-regulatory/research-development/scientific-guidelines/clinical-efficacy-safety/biostatistics. For instance, the topic of adaptive clinical trials has been covered in the 2007 reflection paper on methodological issues in confirmatory clinical trials planned with an adaptive design [36].

Looking beyond biostatistics, EMA's scientific guidelines on clinical evaluation of medicines for specific disease areas are of great value. Suppose one works on clinical trials for the central nervous system disorders. By exploring the link https://www.ema.europa.eu/en/human-regulatory/research-development/scientific-guidelines/clinical-efficacy-safety/clinical-efficacy-safety-nervous-system one can find guidance documents that describe clinical development strategies for treatment of migraine, multiple sclerosis, depression, etc. Likewise, one may want to search the FDA website for disease-specific clinical development guidelines. Your clinical colleagues could be pleasantly surprised with the value you can add by leveraging this knowledge for them!

Knowledge of industry-wide standards such as CDISC (Clinical Data Interchange Standards Consortium) and MedDRA (Medical Dictionary for Regulatory Activities) are also important; however the depth of this knowledge would have to be personalized.

4.4.3 Understanding of the clinical trial protocol

The importance of the clinical trial protocol cannot be understated. This document is central to any clinical trial because it defines the rules according to which the trial should be conducted. Carefully designed and implemented trial protocols would lead to credible results, and vice versa: protocol violations can introduce bias and nullify all efforts of the entire clinical research team.

The pharma industry has well-established practices for clinical trial protocols. It is one of the biostatistician's major responsibilities to contribute to the protocol development and implementation. Essentially, any protocol describes three major components: trial design, trial conduct, and data analysis. These components are pre-specified to reflect scientific considerations underlying the trial and to ensure objective analysis of the collected experimental data. The reporting of the trial results is another important ingredient, but it is a separate task which is outside of the scope of the trial protocol. Understanding of basic elements of the trial protocol is an absolute "must" for every biostatistician, and we shall now elaborate on the specifics.

Once again, our best assistant is the ICH E9 guidance [61]. First and foremost, for any clinical trial it is important to have an understanding of the "big picture"—the clinical development program and the purpose of the current trial in this context. As such, it is always very helpful to have other documents at hand, such as the clinical development plan (CDP), the target product profile (TPP), the integrated development plan (IDP), etc. There are two distinct types of clinical trials—exploratory and confirmatory. Exploratory (phase I–II) trials are designed to generate new clinical hypotheses, and confirmatory (phase III) trials are designed to test these specific hypotheses.

For any clinical trial in the development program, the core element is the clearly formulated *study objectives*. The objectives should include the research questions to be answered in a given trial. Many trials pursue multiple objectives and the investigators should rank these objectives in the order of importance (e.g. primary, secondary, or exploratory). After the trial objectives have been quantified and ranked in the order of their importance, an investigator should choose a study design to achieve these objectives. The following considerations are important [59, 61]:

- *Type of design:* parallel group, crossover, factorial, dose escalation, dose ranging, enrichment, titration, group sequential, or adaptive design.

- *Primary research hypothesis* (in the case of confirmatory trials): test for equality, superiority, non-inferiority, or equivalence.

- *Study endpoints (response variables):* normal, binary, count, or time-to-event. The sample size justification will be driven by the primary endpoint.

- *Control group:* placebo control, active control, or multiple controls.

- *Study population:* normal healthy volunteers, patients, or special population (e.g. pediatric or geriatric patients).

- *Additional considerations:* measures to protect the design from bias (randomization, blinding, allocation concealment).

- *Statistical methodologies* for data analysis.

Perhaps one of the most difficult questions to be addressed by a statistician is justification of the sample size. The question "How many subjects do we need in the trial and what is the power?" is frequently the first one asked by the clinician; yet, this is probably the very last piece of information to be included into any trial protocol.

A scientifically sound clinical trial design should clearly state statistical criteria for sample size planning. These statistical criteria stem from the trial objectives and may include: statistical power, estimation precision, probability of (correct) go/no-go decisions, assurance, etc. It is important to have consistency among trial objectives, trial design, statistical analysis, and sample size. O'Hagan and Stevens [76] proposed the following insightful approach to sample size determination. Given the trial design and statistical criteria, define: 1) The *analysis objective*—what we hope to achieve as a positive outcome of the trial (e.g. decision and/or inference). 2) The *design objective*—how sure we wish to be of achieving the analysis objective. Then the sample size determination follows a two-step procedure. First, identify what data would lead to achieving the analysis objective. Second, select the sample size to give the design objective's required assurance of obtaining data that would achieve the analysis objective.

For instance, suppose we are designing a confirmatory trial to demonstrate superiority of the new drug over a placebo. The analysis objective is formulated as: "To reject the null hypothesis of no treatment difference at the 5% significance level (one-sided P-value < 0.05)". Therefore, for the design objective, we would require that the study sample size n should be such that there is, say, 90% probability of achieving the analysis objective (one-sided P-value < 0.05), if the true mean treatment difference equals some clinically meaningful value $\Delta > 0$.

We should be mindful about the text for sample size justification that goes into the study protocol. Unfortunately, it is not uncommon to have many instances of wrong statements, such as: "The design with n subjects will provide 90% power to detect a mean difference of Δ between two treatment groups." In fact, the intended statement is "The design with n subjects will provide 90% power *to detect statistically significant difference between two treatment groups (at a given significance level)* when the true mean difference is Δ." Such subtle details require care and reinforce the importance of a biostatistician's engagement in the development of any clinical trial protocol.

Finally, here is a simple hint how to get some good examples of clinical trial protocols. Visit a website of a major medical journal, such as *The New England Journal of Medicine*, and search for recent publications of the results of a randomized controlled clinical trial in the disease area of interest. Download the paper, and look for the "Supplemental Materials", which should include the clinical trial protocol as one of the appendices. Chances are very high that you will find the protocol (and the statistical analysis plan as a bonus!) Given that the paper has been published in such a reputable journal, the protocol merits reading and can potentially be a very useful learning example.

4.4.4 Knowledge of statistical methodologies for protocol development

It is impossible (and unnecessary) to cover all statistical methodologies that may be useful for protocol development. The toolkit will be determined by one's background (cf. §4.3), therapeutic area, regulatory considerations (many of which will be very useful and some of which will be restrictive), and established best practices in the industry. What we should keep in mind is the main purpose of any clinical trial—to obtain reliable answer(s) to the question(s) of interest. Having this perspective, a solution can be found for any particular trial protocol.

Let us start with the design part. As mentioned by Grieve [52], the three most important areas where statisticians can contribute and be influential are *design, design and design* (= D^3). The three fundamental techniques in the clinical trial design are randomization, blinding, and choice of control group. The purpose of these techniques is to minimize potential for bias in the study. An exemplary design for clinical investigation is randomized, double-blind, placebo- or active-controlled clinical trial. When properly implemented, it can generate credible and reliable results.

Randomization refers to generation of a sequence of treatment assignments by means of some known random mechanism. The key merits of randomization in clinical trials are (at least) four-fold: randomization helps mitigate various experimental biases (such as selection bias); it promotes "comparability" of treatment groups at baseline; it contributes to the validity of model-based statistical tests; and it can form the basis for randomization-based inference. The number of available randomization techniques is huge; see, for example [88]. Theoretical knowledge of these techniques and experience with applying them in practice is a big asset. Randomization-based inference (through Monte-Carlo re-randomization tests) can be easily done in seconds with modern computing powers [89]. These tests can be supplemental to parametric (model-based) tests, and under some circumstances these re-randomization tests provide more robust and reliable results. In addition, randomization-based inference ensures a consistency link between the trial design and subsequent data analysis.

The *blinding* (or masking) of treatment assignments is not a statistical method per se, but a process to ensure that treatment assignments are not known or easily ascertained by the patient, the investigator, and the outcome assessor. The ICH E9 guideline distinguishes *open-label* trials in which "the identity of treatment is known to all", *single-blind* trials in which "the investigator and/or his staff are aware of the treatment but the subject is not, or vice versa"; and *double-blind* trials in which "neither the study subjects nor any of the investigator or sponsor staff who are involved in the treatment or clinical evaluation of the subjects are aware of the treatments received". The double-blind study is regarded as the optimal approach, and if it is infeasible, a single-blind study with appropriate measures for bias mitigation is suggested. The existence of the problem of potential bias due to lack of (or improper) blinding in clinical trials is well documented and some useful statistical techniques for handling these issues are available [8, 9].

The use of a *control group* as the reference for treatment comparisons is another fundamental trial design aspect. The main purpose of the control group is to facilitate discrimination of patient outcomes caused by the test treatment from outcomes caused by other factors (e.g. natural progression of the disease, observer or patient expectations, or other treatment). The importance of this topic is signified by the presence of the ICH E10 guideline [60] that provides a comprehensive review of different types of controls in clinical trials, and outlines important issues (with numerous implications for statisticians), such as non-inferiority and equivalence trials, active-controlled trials, assay sensitivity, etc. While these issues may not frequently pop up in the work of many biostatisticians, the acknowledgement of their existence and the reference to ICH E10 is certainly important.

Another design consideration which is (again) worth mentioning is the sample size determination. The sample size cannot be too small because the study may be underpowered and may fail to detect statistically significant and clinically relevant treatment difference. At the same time, the sample size cannot be too large because it can yield significant results that are of no clinical relevance (in large and very expensive studies). Various statistical methods for sample size determination are available in many statistics books [21, 64]. A "guiding principle" is to have a clearly formulated research question, carefully investigate the literature (using meta-analysis, if appropriate), consult with the clinical trial team (medical doctors/clinical pharmacologists) to elicit the values of parameters for sample size planning under the "base case" scenario, and perform sensitivity analyses to ensure the proposed sample size fulfills the study objectives with high probability. Proper documentation of the sample size determination (and reproducibility of the calculations) is mandatory.

As a follow-up, let us consider the conduct of the clinical trial. Many modern clinical trials utilize statistical monitoring of data and may have one or more pre-planned interim analysis. The interim analysis (IA) is defined as "*any analysis intended to compare treatment arms with respect to efficacy or safety at any time prior to the formal completion of the trial*" [61]. The purpose(s) of an IA may be: i) early stopping for overwhelming efficacy; ii) early stopping for futility; iii) stopping for safety reasons; and iv) modifying the study design (e.g. sample size reassessment; dropping an arm; enrichment of a subpopulation, etc.) Regardless of the purpose, the IA must be carefully planned for in the protocol. The IA poses many statistical challenges, such as control of the error rates; the fact that interim data can be immature, highly variable and fluctuating over time; and the fact that dissemination of the IA results can introduce bias and compromise integrity of the ongoing study. Without delving into much details here, one can derive a lot of useful and practical knowledge from studying group sequential design (GSD) methodologies [63] and principles of data monitoring committees (DMCs) [33]. The one-sentence message about IAs is that they may indeed help expedite clinical development and benefit patients; however they are neither free nor a means to satisfy investigators' curiosity, and biostatisticians should play a pivotal role in both planning and execution of IAs.

Let us now discuss the data analysis part, which is (rightfully) considered to be the "bread and butter" of any biostatistician's work. The data analysis actually starts with specification of the analysis sets in the study protocol. This is a crucial step which sometimes does not receive due attention. Note that even a flawlessly designed and executed clinical trial may still yield biased results, if the analysis is performed on a selective set of patients (e.g. non-intention-to-treat analysis populations). Definitions of the full analysis set, per protocol set, safety set, etc. should be explicit in the "Data Analysis and Statistical Methods" sections of the protocol.

Definition of the primary endpoint(s) and specification of statistical methods for analyzing them are the next step. The primary analysis usually involves some kind of regression modeling (e.g. linear models for normally distributed outcomes; generalized linear models for binary or categorical outcomes; or survival models for time-to-event outcomes). The model should account for important covariates (some of which may have been used as stratification variables at the randomization step), and for possible repeated measurement data structures (e.g. in case of crossover designs). Also, some primary outcome measures are commonly transformed before the analysis (e.g. using log-transformation for Cmax or AUC in clinical pharmacology studies). A person who has taken courses in linear regression and generalized linear models (cf. §4.3) would definitely find this part enjoyable and entertaining!

In practice, model assumptions can be violated and data from some trial participants may be missing. The approaches to handling missing data, the sensitivity analysis and the supportive analysis strategy should be pre-specified in the protocol as well. Ideally, we should have consistent results/conclusions from the primary and supportive analyses. All these considerations lead us to the concept of *estimands*, which we will discuss in more detail in §4.5.

Statistical methods for analysis of safety data (which is often not the primary objective) merit a separate (book-length) discussion [19]. In brief, we face numerous statistical challenges when analyzing safety data, primarily due to the fact that individual trials are limited in size and evaluation of drug safety requires examination of data from multiple heterogeneous sources (meta-analysis). Industry best practices for safety data analysis are still evolving.

Exploratory data analysis is another important ingredient of clinical trial protocols. This often includes analysis of biomarker data, modeling of PK/PD relationships, and use of graphical tools to explore important trends. Various data science and machine learning tools are useful in this regard and are emerging.

Note that there will always be disease-specific statistical tools, such as survival analysis in oncology, negative binomial regression for modeling lesion count data in multiple sclerosis, or joint longitudinal and time-to-event modeling in studies of HIV/AIDS. The statistical toolkit will most likely evolve over time. However, the philosophy will remain the same: the goal is to extract credible and reliable results/conclusions from collected experimental data.

4.4.5   Statistical software

Modern applied statistics is incomplete without software for implementing its state-of-the-art methodologies. With rapid technological advances, one can expect that statistical software will evolve and become more and more elaborate; however the main principles will remain the same. Section 5.8 of ICH E9 [61] has the following statement: *"The computer software used for data management and statistical analysis should be reliable, and documentation of appropriate software testing procedures should be available."* This statement highlights the importance of principles of reproducible research and formulates key requirements for the statistical software. In an effort to be pragmatic, we define the following four categories of statistical software: 1) software for clinical trial design; 2) software for clinical trial data analysis; 3) software for visualization/reporting of clinical trial results; and 4) software whose existence should be recognized.

In the first category we naturally have software for sample size calculation and simulation of clinical trial designs. Many pharma companies for this purpose use their own preferred set of commercial validated software packages, such as nQuery, PASS, EAST, etc.

These packages are menu-driven and very easy to run. The advantage is that the output is obtained without writing programs, and therefore does not require validation/double programming. Obtaining a screenshot of an individual window with the output (using 'Print Screen') and pasting the clipboard content into a separate MS Word document are sufficient for producing documentation. It would still require some kind of QC by another statistician (e.g. checking the assumptions and appropriateness of a statistical method to calculate sample size/power, checking the parameter values, etc.) However, when these packages are available, one should definitely take advantage of them because they can save a lot of time and effort.

In many clinical trials, the planned primary analysis uses complex statistical methodologies (e.g. mixed-effects model with repeated measures (MMRM), generalized estimated equation (GEE) models, etc.) for which limited or no analytical formulas for sample size are available. Also, some trials use complex adaptive designs with interim decision rules—most likely there are no closed form expressions for sample size and power calculations. In such cases a useful approach is to use Monte Carlo simulation. This leads us to one of the most powerful weapons in the biostatistician's arsenal—the R software (https://www.r-project.org). Hands-on knowledge of R significantly increases the value of any statistician; in fact the "experience in R programming" becomes increasingly explicit in many job descriptions. The main advantage of R is two-fold: 1) it is free; and 2) its capabilities are constantly expanding through user-created packages which implement statistical techniques, scientific graphics, etc. Biostatisticians can do wonders with R. Just one quick example: a line of code in R—sample(x=c(rep(0,25),rep(1,25)), size=50, replace=FALSE)—generates a randomization list (a random sequence of treatment assignments, 25 for each of the two treatment groups). For the second author, R has been a software of choice for more than 15 years. Monte Carlo simulations of complex adaptive design procedures (with potential non-convergence of algorithms at some simulation runs due to the random nature of the data) is an ideal setting for application of R programming. A recent trend in the industry is to use R Shiny package (https://shiny.rstudio.com/) for creating interactive web-based apps straight from R. This is a perfect opportunity for biostatisticians to show their value to clinical colleagues!

For data analysis, the industry standard is SAS. We already discussed SAS statistical programming aspects in Chapter 2—all of them are applicable here. We just add a few notes on statistical analysis procedures. If one has to name one most influential procedure for data analysis in SAS, this would probably be PROC MIXED. If properly implemented, it can be used to perform any kind of linear regression modeling on responses that are (approximately) normally distributed (perhaps after transformation), including linear mixed effect models (such as MMRM). Useful analogues of PROC MIXED are PROC GLIMMIX for generalized linear models and PROC NLMIXED for nonlinear models. The art of modeling involves careful specification of fixed effects (e.g. treatment, visit, treatment-by-visit interaction, and important prognostic covariates at baseline), possible random effects and covariance structures to account for multiple correlated measurements within each subject. An unstructured covariance matrix (UN) is frequently used for the primary analysis. However, this analysis may fail to converge, and additional analyses with other covariance structures should be planned for. The survival analysis in SAS is facilitated by three main procedures: PROC LIFETEST (for nonparametric methods), PROC LIFEREG (for parametric models), and PROC PHREG (for the Cox proportional hazards model). Many more useful procedures for data analysis are available in SAS, and their application will certainly be context-specific.

For visualization/reporting of clinical trial results, there are at least two choices. The first choice is Microsoft Excel. Once data have been analyzed and the desired results (say, estimates with confidence intervals) are available, these results can be exported into Excel, and from there one can generate high quality graphics. In fact, some clinical journals require that figures in the publications are generated this way—to ensure consistent standards for the journal. The second choice is R. As we mentioned, capabilities of R are constantly growing, and today we have excellent tools to produce scientific graphs. For instance, the R package ggplot2 (https://cran.r-project.org/package=ggplot2) can be used to generate practically any graph that is relevant to the pharma industry. Numerous tutorials on ggplot2 are available online and one can master this technology relatively quickly. Of course, there are many other ways to generate graphs—SAS has now significantly improved graphical capabilities; fancy users may decide to go with the grand master Tableau Software (https://www.tableau.com/) or prefer some other technologies. However, these are definitely beyond the "minimal sufficient" set of tools we are trying to elicit here.

Finally, let us briefly discuss software whose existence should be recognized. One of the best biostatistician's allies in the big pharma industry is the pharmacometrician [67]. This individual is an expert in PK/PD modeling and simulations and he/she plays a major role in model-based drug development. Pharmacometricians use their own set of tools which may be a puzzle for statisticians; for instance, the NONMEM (for NONlinear Mixed Effects Modeling) software, developed by Stuart Beal and Lewis Shiner at the University of California, San Francisco, in the late 1970's. NONMEM is perhaps as important for pharmacometricians as SAS is for statisticians. It is very intricate and complex for non-specialists; yet it is a time-proven powerful tool to do modeling and simulations for calibration of dose/regimen schedules. Another useful and increasingly popular software for pharmacometrics is Monolix—http://lixoft.com/products/monolix/. Importantly, it has a free license for academic institutions.

4.4.6 Communication skills

"Excellent communication skills", "ability to explain complex statistical concepts in simple terms to non-statistical colleagues", "excellent stakeholder management", "strong negotiating skills"—these are increasingly common requirements from biostatistician job descriptions. Drug development is a team effort, and excellent communication skills is indeed essential. In school, we are taught that copying is bad and prohibitive and we should excel as individuals; in the pharma industry, copying is often necessary for success. Given the global nature of drug development and increasingly complex structures of big pharma companies, efficient communication becomes a major prerequisite for business success. Quoting a former American Statistical Association (ASA) President Nancy Geller [47]: "*In statistics the three most important things are communication, communication, communication!*" $(= C^3)$.

The first step for anyone who starts working on a clinical trial/project would be to do some kind of "stakeholder mapping"—identifying key individuals with whom one should have business interaction and communication. For a biostatistician, this map will likely be multidimensional and include:

- *Clinical* trial/project team members: clinical trial leader, medical director, PK scientist, regulatory scientist, project manager, etc.

- *Technical* team members (within the company): statistical programmer, data manager, pharmacometrician, bioinformatician, etc.

- *Outsourcing* team members: CRO statistician, CRO programmer, etc.

- *Line function management*: manager, one level over manager, and direct reports (if any).

Within each dimension, it is always good to establish "rules of engagement"—to define common goals, expectations, and commitments. In practice, it is common to have regular 1–1 meetings with key stakeholders (at least with the manager and direct reports). Simply, efficient communication means that everyone is up-to-date with ongoing business activities; the timelines and deliverables are well-defined; the progress is carefully monitored; and the work gets done with high quality, in time, and in compliance with the company and regulatory standards. It is difficult to come up with a universal solution for how this can be set up—it has to be personalized. The criteria for efficient communication are part of many pharma companies "Performance and Behaviors" metrics, and these vary across the industry. However, the essence (why being an efficient communicator is important for any biostatistician) is captured in the following quote [12]:

> "Statistical results are of little value if the client doesn't understand them and can't put them to work. The success of an industrial statistician is a direct function of the impact of his or her work on company business."

4.4.7 Knowledge of processes

As we discussed in §4.2, business operating models can be quite complex and multi-dimensional. Understanding of the environment (organizational structure, business processes, and company SOPs) is certainly a must. With no claim to ultimate truth, here are some personal thoughts/recommendations in this regard:

- Learn the foundational aspects of your company—its history, its mission, its vision, its core principles, its portfolio, and its leaders.

- Learn the structure of your company—its major business units, departments, divisions, research groups and their interplay, and understand your role and purpose in this system. Consider the following analogy: you are in the middle of the universe. You can look in the microscope and see (at different levels) the cell, the atom, the nucleus, the electron, etc. You can also look in the telescope and see (at different levels) the planets, the stars, the galaxies, etc. All these elements serve their purpose in a very complex universal structure.

- Have at hand your job description and your company's onboarding plan—it should be a useful reference of what the expectations are and what you need to do in order to succeed in your current role and beyond.

- Obtain access and bookmark links to all systems you need for doing business. These include various general portals (company SOPs, time tracking systems, etc.), technical portals (data repositories, programming environments, document management systems, etc.), and SharePoint sites for different teams and sub-teams you may be part of. Keep in mind that these may change over time—some systems may retire and new systems may come into play.

- Keep track of the work you are doing throughout the year and document all accomplishments (including small ones). This information will be needed for mid-year and end-of-year performance evaluation, which will eventually determine your compensation, career development, etc.

- Create and maintain your development plan—where you stand currently and what you want to achieve in the (near) future. Make sure you discuss this plan periodically with your manager—it is one of his/her objectives to ensure development, growth, and retention of talent (you).

We finalize our "minimal sufficient" set with one more ingredient—attitude toward work. Holding a positive attitude and the desire for constant learning and improvement are essential.

4.5   Advanced biostatistics toolkit

In this section we discuss some advanced statistical methodologies that are hot topics in modern drug development. They represent steady trends that are likely to continue and not fade out in the near future. Please note that this is by no means an exhaustive list; neither should it be viewed as a "must learn" set of topics. This selection primarily reflects the second author's experience and research interests. It can be also viewed as a "navigation guidance to selection" that can save a lot of time if a reader encounters some problems in these areas and wants a quick reference. Our (subjective) selection is as follows:

- Adaptive designs (§4.5.1).

- Basket, umbrella, platform trials and master protocols (§4.5.2).

- Dose-finding methods (§4.5.3).

- Multiplicity issues (§4.5.4).

- Estimands (§4.5.5).

Phase	Adaptive Design	Benefits	Cautions/Limitations
I	Dose escalation	Flexible; can incorporate historical data via prior distributions (Bayesian); can identify MTD and other parameters of interest more precisely and with lower sample size than SAD/MAD dose escalation designs	Sensitive to choice of prior distribution and model assumptions; careful calibration via simulations is required
II	Adaptive dose ranging	Promising when the initial design is not good and/or interim parameter estimates have low variability; can increase precision in estimating dose-response curve	Outcome must be a short-term endpoint; collected data should be quickly available for interim analysis and design adaptations
II	Adaptive randomization	Addresses ethical concerns (more patients allocated to the empirically better treatment); can increase statistical efficiency (precision/power)	Cautioned by the FDA in the confirmatory setting
II/III	Seamless phase II/III	Avoidance of administrative wait between phase II and III protocol activation; can help reduce the development time	"Firewall charter" must be in place to maintain trial integrity; proper statistical adjustments are required to account for interim selection / combination of data from phase II and III
III	Group sequential design	Enables early stopping for efficacy, futility, or harm; viewed as "well understood" by the FDA	Cannot alter maximum sample size or events in an unblinded manner; uses nonconventional parameter estimates and confidence intervals
III	Sample size reassessment	Can help mitigate impact of incorrect assumptions on study power	Unblinded SSR requires strict firewalls to prevent leakage of interim information; uses nonconventional parameter estimates and confidence intervals
III	Subpopulation selection	Can eliminate nonperforming subgroup(s) and focus on a subpopulation that is more likely to benefit from treatment	Sample size for the subpopulation may need to be increased; requires careful calibration via simulation and special statistical analysis approaches

FIGURE 4.3
Major types of adaptive designs, their benefits and limitations.

- Quantitative decision-making support (§4.5.6).

- Digital development (§4.5.7).

4.5.1 Adaptive designs

One may argue that adaptive designs has been one of the hottest topics in clinical trial biostatistics over the past 25 years or more. Numerous published research methodology papers, M.Sc. and Ph.D. dissertations, special issues of statistical journals, scientific conferences, books, and even regulatory guidelines on the subject signify the importance of adaptive designs in drug development. The usefulness of the concept of "adaptivity" can be traced to Charles Darwin (1809–1882) who stated that *"It is not the strongest of the species that survives, nor the most intelligent survives. It is the one that is most adaptable to change."* The interested reader is referred to three recent papers: [6, 15, 80] which provide excellent background on adaptive designs and their role in clinical trials.

Without delving into technical details, let us answer a basic question: "Why should one consider adaptive designs?" The matter is that at the planning stage of a clinical trial various assumptions must be made, such as the choice of the value of a clinically relevant treatment effect, the value of the primary outcome variance, the dropout rate, etc. Naturally, there is uncertainty in these assumptions, and the earlier the stage of development, the higher the uncertainty. Inaccurate assumptions increase the risk of trial failure. Having an option to modify trial design according to carefully pre-defined rules based on accumulating experimental data (adaptive design) is an attractive idea—uncertainty can be reduced and trial objectives can be accomplished with greater confidence and, potentially, in a shorter timeframe compared to a fixed (non-adaptive) design. As such, one could envision various potential benefits of adaptive designs. In the exploratory setting, adaptive dose-finding trials can lead to more accurate determination of the MTD, and adaptive dose-ranging trials can lead to more accurate characterization of the dose–response curve and increased chances for taking the right dose(s) into confirmatory trials. In the confirmatory setting, group sequential designs can expedite business decisions by stopping trials early for overwhelming signals of efficacy, futility, or safety; adaptive enrichment designs can eliminate nonperforming subgroups of patients at interim analyses (and focus on the subpopulation(s) that is/are most likely benefit from treatment); and sample size reassessment designs can alleviate incorrect assumptions on the outcome variance and/or the treatment effect.

Given potential promises of adaptive designs, one may ask a question: "Should all clinical trials be adaptive?" Of course, nothing comes for free. Indeed we have several potential challenges with adaptive designs. First, there are statistical challenges: design adaptations may introduce bias into the data: conventional statistical inferences on the treatment effect (point estimates, confidence intervals and P-values) may not be reliable anymore. Second, there are operational challenges: adaptive trials require more upfront planning compared to traditional designs; robust infrastructure to implement design adaptations must be in place; unblinding of interim data must be done with care to avoid operational bias; and speed of recruitment does matter (the faster subjects are recruited into the study, the less opportunity there is for adaptation). Third, there are regulatory challenges: while regulators usually welcome adaptive designs in exploratory trials, they review proposals with confirmatory adaptive designs with higher scrutiny due to limited experience and concerns about control of the type I error rate.

One can be assured that potential benefits of adaptive designs outweigh their limitations. For instance, Hatfield et al. [55] performed a systematic review of registered clinical trials and found that uptake of adaptive designs in clinical drug development is actually gaining traction and increasing. Oncology remains the main therapeutic area for applications; however other disease areas, such as mental health, musculoskeletal, gastroenterology, etc. have seen increased application of adaptive designs. The aforementioned review excluded phase I trials, and therefore the majority of adaptive design applications were found to be in phase II. The group sequential design was found to be the most popular adaptive design methodology for phase III.

Figure 4.3 gives a brief summary of various adaptive designs, their benefits, and limitations. Overall, it is fair to say that adaptive designs (when properly implemented) can improve flexibility and efficiency of clinical trials. The field of adaptive trials is expanding. The key regulatory references on adaptive designs are:

- 2007 EMA reflection paper on methodological issues in confirmatory clinical trials planned with an adaptive design [36].

- 2016 FDA guidance on adaptive designs for medical device clinical studies [42].

- 2018 FDA (draft) guidance on adaptive design clinical trials for drugs and biologics [45] (which is the updated version of the earlier 2010 draft).

In the future, continuation of statistical research on adaptive design methodologies and wider acceptance of adaptive trials among stakeholders involved in clinical R&D can be expected.

4.5.2   Basket, umbrella, platform trials and master protocols

Master protocols is a novel and very useful concept in drug development. Modern clinical research is increasingly focused around the idea of developing personalized medicine (finding the right treatment for the right patient at the right time). This is particularly important in oncology where carcinogenesis and treatment are closely related to molecular and genetic mutations. Developing treatments for personalized medicine is a big challenge, because the number of potential options (different therapies and their combinations) may be huge and the patient populations may be small. There is a strong need for clever designs to evaluate multiple therapies across a spectrum of indications in heterogeneous patient populations.

A *master protocol* is defined as an overarching protocol designed to answer multiple research questions within the same overall trial infrastructure [105]. There are 3 types of master protocols: 1) umbrella trial; 2) basket trial; and 3) platform trial.

The *umbrella* trial evaluates multiple targeting agents, generally within a single tumor histology (e.g. non-small cell lung cancer). The research question is "Which drug has the most impressive therapeutic index to pursue in further development for the given indication?" Thus, umbrella designs can be thought of as "compound finder" trials. Since multiple drugs are tested, umbrella trials require multi-stakeholder collaboration (e.g. different sponsors or different development teams within an organization).

The *basket* trial evaluates a single targeted agent across multiple indications (e.g. breast, ovarian, pancreatic, etc.) The research question is "Which indication exhibits the most impressive therapeutic benefit for the given drug?" Thus, basket trials can be thought of as "indication finding". Unlike umbrella trials, basket trials may not require coordination among different sponsors.

The *platform* trial is the most complex one [93]. It involves multiple treatment arms: either a fixed number (e.g. a trial with K treatments from different sponsors), or an adaptive number (due to the possibility of adding/dropping arms). In addition, it involves multiple biomarker strata, and some new strata may be added once a biomarker assay and an investigational targeted drug become available for the trial. Finally, the design may use either fixed randomization or response-adaptive randomization (RAR) to dynamically skew allocation of new patients to the empirically better treatment groups within biomarker profiles. The research question is "Which successful compound/biomarker pairs should be taken further in development to small, more focused confirmatory phase trials?" Platform trials may be most appropriate in phase II, based on short-term endpoints as surrogates for clinical response.

One famous application of a platform trial design is the BATTLE trial (Biomarker-Integrated Approaches of Targeted Therapy for Lung Cancer Elimination) [65, 108]. This was a prospective, biopsy-mandated biomarker-based response-adaptive randomized study in patients with non-small cell lung cancer. Upon biopsy, eligible patient tissue samples were analyzed for biomarker profiling and classified into 5 biomarker strata. Within each stratum, patients were randomized among four targeted therapies. The study objective was to identify subgroups of patients who are most likely to benefit from specific agent(s). The primary endpoint was the 8-week disease control rate (DCR) (a good surrogate for overall survival in this indication), which made RAR feasible in this study. Between November 2006 and October 2009, 341 patients were enrolled, and 255 patients were randomized (the first 97 using equal randomization, and subsequent 158 using RAR). Final group sizes were unequal across the treatments (due to RAR), as was the prevalence of the biomarker strata. There were several interesting findings; e.g. in the KRAS or BRAF mutation-positive patients, the drug sorafenib showed very promising activity: 8-week DCR=79% (vs. historical 30%). Of course, due to small sample sizes, the results had to be interpreted with caution and the study was considered as "hypothesis generating". However, it provided impetus for further investigation (BATTLE-2 study [82]), which is already another story...

Recognizing the importance of the topic, in September 2018 the FDA issued a draft guidance "Master protocols: Efficient clinical trial design strategies to expedite development of oncology drugs and biologics." [46] We anticipate continuing interest in this topic, the development of new statistical methodologies (using Bayesian methods), comparisons of existing designs by simulation and identifying best ones for use in practice, and development of information technology platforms to facilitate global implementation of these designs.

4.5.3   Dose-finding methods

Appropriate dose finding and dose selection are foundational to the development of drugs with favorable benefit/risk properties. This was acknowledged long ago by Paracelsus (1493–1541): *"All substances are poisons; there is none which is not a poison. The right dose differentiates a poison and a remedy."* The importance of appropriate dose selection and its optimization is also reflected in ICH E4, E8, and E9 guidelines which became effective in the mid-1990's and onwards. In the 2000's, lead experts from the pharma industry formed a PhRMA adaptive dose-ranging studies working group (ADRS WG) which provided evaluation and recommendation on innovative dose-ranging designs and methods [14, 31]. Continuing in this spirit, statistical methodologists from Novartis requested a CHMP qualification opinion on the MCP-Mod (Multiple Comparison Procedure–Modeling) methodology, originally developed by Bretz et al. [17]. In 2014, the CHMP adopted the MCP-Mod approach as one that *"...will promote better trial designs incorporating a wider dose range and increased number of dose levels..."* [37]. This was the first statistical methodology which received EMA's qualification opinion.

Needless to say, knowledge of state-of-the-art dose-finding methodologies and ability to apply them in practice significantly increase the value of any biostatistician. The innovation trend for dose-finding will certainly continue and we see at least two distinct areas in this regard.

One area is adaptive dose-finding methods in phase I and phase I/II trials, especially in the areas of oncology and immuno-oncology. Statistical research in the 1990's led to introduction of many novel adaptive dose escalation designs for phase I oncology trials, such as the continual reassessment method (CRM) [78], the escalation with overdose control (EWOC) [5], Bayesian decision-theoretic approaches [101], etc. Some of these methods have become standard in many pharma companies, replacing the good old 3+3 design; however, these designs were developed for finding MTDs of cytotoxic anticancer agents (for which an implicit assumption is that both the probability of toxicity and the probability of therapeutic response are increasing with dose). The discovery and development of targeted therapies require different design considerations, because targeted therapies are much less toxic and the efficacy response may reach its peak at doses way below the MTD. Seamless phase I/II designs have been introduced to identify dose(s) that exhibit promising signals of efficacy (phase II goal) and have acceptable safety (phase I goal) in a single study [106]. Phase I/II trials would make more efficient use of important safety and early efficacy (biomarker) data, and would avoid the administrative wait between phase I and II protocol activation. One nice example of a phase I/II trial is a *cohort expansion* design, where phase I toxicity-based dose escalation is followed by a phase II single-arm or multi-arm randomized selection design using efficacy response. In August 2018, the US FDA issued a draft guidance "Expansion cohorts: Use in first-in-human clinical trials to expedite development of oncology drugs and biologics" [40]—this gives us assurance that regulators recognize the importance of this topic and encourage sponsors to consider these designs in practice.

Another area is phase II dose-ranging trials which are driven by the primary goal of estimating the dose–efficacy response and identifying dose(s)/regimen(s) that are most promising for confirmatory testing in phase III. Phase II dose-ranging trials are usually randomized, placebo- and/or active-controlled, parallel group designs with sample sizes up to several hundred patients. In reality, there is uncertainty about the dose–response relationship, and there are numerous questions regarding how to optimally design such studies (e.g. how to choose the number of doses, the spacing between the doses, the allocation ratio, and the total sample size). The MCP-Mod methodology was originally developed for normal response data, collected at a single time point. Subsequently it was extended to scenarios with binary and count data, longitudinal responses, adaptive two-stage procedures, etc. Further extensions of MCP-Mod are expected in the near future; e.g. to incorporate exposure-response models. Also, establishment of industry best practices, say, understanding of which indications and which types of investigational drugs the MCP-Mod methodology is best suited for is still evolving. For further details, refer to the edited volume by O'Quigley et al. [77]—its part III provides a nice account of the current body of knowledge on MCP-Mod and some other useful techniques for dose-ranging trials.

4.5.4 Multiplicity issues

A textbook example of a clinical trial—randomized, double-blind, placebo-controlled design comparing efficacy of a new intervention vs. placebo via testing a single primary research hypothesis—may be of limited use in today's clinical research. Many clinical trials are designed with an attempt to answer multiple research questions, while maintaining scientific validity as defined in the regulatory guidelines. When multiple statistical tests are performed within the same study, appropriate adjustments are necessary; otherwise the results will have exploratory nature and may not be acceptable in a broader scientific community. From a statistics course, we know that if we test K research hypotheses (each at significance level α), then the probability of making at least one false positive finding is $P = 1 - (1 - \alpha)^K$. For instance, if $K = 10$ and $\alpha = 0.05$, then $P \approx 0.40$. We can make the Bonferroni adjustment and test each hypothesis at the significance level α/K. Then the probability of making at least one false positive finding becomes $P = 1 - (1 - \alpha/K)^K$, and we can show using simple calculus that $P \leq \alpha$. However, what if, in fact, K is large? If we go with the Bonferroni approach, we can be very conservative and not be able to reject individual null hypotheses (each at the level α/K) unless the treatment effect is overwhelming. The major regulatory prerequisite for a confirmatory trial is the *strong* control of the familywise error rate, which means that the probability of making at least one false positive finding should be at most α for *any* configuration of true and non-true null hypotheses in the family. Are there methods more powerful than the Bonferroni approach which also ensure the strong control of the familywise error rate? The answer is "yes", and these methods have been studied extensively over the past years.

A succinct summary of all statistical methodologies that are useful to address multiplicity issues in clinical trials is beyond the scope of the current chapter. As with adaptive designs, numerous research papers, special journal issues, books, and regulatory guidelines have been written on this topic. The field is constantly evolving and new methodologies are being developed by lead experts in this field; see, for instance, [30]. Two most recent regulatory guidelines the reader should definitely get familiar with are:

- 2017 EMA draft guideline on multiplicity issues in clinical trials [35] (which is a revision of the original 2002 version).

- 2017 FDA draft guidance on multiple endpoints in clinical trials [41].

Here we just acknowledge different sources of multiplicity that may arise in modern clinical research:

- *Multiple treatment arms.* Phase II dose-ranging trials (and some phase III trials) involve several active dose groups and the placebo group and call for estimation/testing of multiple contrasts (active vs. placebo).

- *Multiple endpoints.* Some clinical trials use co-primary endpoints (e.g. progression-free survival and overall survival in oncology; measures of cognitive impairment and functional performance in trials of Alzheimer's disease).

- *Multiple subgroups.* There may be multiple strata (e.g. defined by different genetic signatures) within which an investigator may wish to perform treatment comparisons.

- *Multiple interim looks.* A classic example is the group sequential design methodology [63].

One can envision complex multiplicity issues that can potentially arise if some of the aforementioned sources of multiplicity are combined together. Are we now convinced that the topic of multiple comparisons belongs in the advanced biostatistics toolkit?

4.5.5   Estimands

The concept of estimands is important and relatively new in clinical development. However, it has been at the forefront of statistics discussions since October 2014, when an expert working group (regulatory and industry statisticians) was formed to develop an addendum to the ubiquitous ICH E9 guideline [61]. In 2017, the draft guideline ICH E9(R1) "Estimands and sensitivity analysis in clinical trials" [58] was released for public comment.

So why did this topic emerge and why is it important? As noted by Mehrotra et al. [73], any regulatory guideline (even such authoritative and well-written as ICH E9) will eventually become incomplete or outdated because of new scientific advances, emergence of some new issues that have not been uncovered before, and lack of specificity on some existing statements. It has been argued that several recent clinical trial protocols and new drug applications lacked logical connectivity between trial design, conduct, analysis and interpretation. A structured framework would therefore help bring transparency in the definition of *estimand* (what is to be estimated in a given trial) and the *sensitivity analyses* (to evaluate robustness of the results to potential limitations of the data and/or departures from the model assumptions).

The ICH E9(R1) proposes the following framework for aligning planning, design, conduct, analysis, and interpretation:

1. *Formulate the trial objectives.* In the context of treatment benefit, the key question is: "What are we trying to estimate—efficacy or effectiveness?" Note a delicate difference: *efficacy* refers to the sole benefit of treatment (the expected effect a patient may reach if he/she takes medication as directed), whereas *effectiveness* takes into account safety/tolerability issues and refers to the treatment effect in the presence of the actual adherence pattern. The choice between efficacy and effectiveness will be determined by the stage of development (e.g. in phase II PoC trials efficacy is more appropriate, whereas in phase III/IV the focus shifts towards effectiveness), the indication (e.g. short-term outcome symptomatic trials will prioritize efficacy, whereas long-term maintenance trials will prioritize effectiveness), and possibly other factors.

2. *Choice of the primary estimand.* Keeping in mind Step 1, define the target of estimation by specifying the population of interest, the variable (endpoint), the way of handling intercurrent events, and the population-level summary of the variable (e.g. in the binary outcome case it may be risk difference, relative risk, or odds ratio). The definition of the primary estimand will determine the main aspects of the trial design, conduct, and analysis.

3. *Choice of the main estimator.* After Step 2 is complete, the statistical methodology for data analysis can be specified. Once the trial data become available, this methodology will be applied to obtain the main estimate. This main estimate will likely be subject to certain assumptions (e.g. due to a selected imputation method for missing data), some of which may be difficult (or impossible) to verify. Therefore, we proceed to Step 4:

4. *Specify the sensitivity analysis.* This step is needed to assess robustness to variations of the assumptions underlying primary analysis (possibly using several different approaches). The sensitivity analyses can enhance the credibility of the trial results.

One obvious advantage of the ICH E9(R1) framework is that it should facilitate a lot of upfront thinking and discussion within clinical trial teams and among other stakeholders. This is an example when the biostatistician (due to his/her unique technical knowledge) can play the role of a scientific leader, gluing together different subject matter expertise. We anticipate that industry-wide discussion on estimands will continue, as will the statistical research on this topic.

4.5.6 Quantitative decision-making support

Many pharmaceutical companies have recognized limitations of the conventional clinical development paradigm when the process is split into discrete phases I, II, and III, each with its own set of research questions and objectives. A recognition of the fact that we should focus not only on individual trials but also on the "big picture" has led to many innovative approaches to optimization of clinical development programs [29] and even portfolios of projects [3].

Around 2012, the second author was fascinated reading a paper by Orloff et al. [79] which gave a perspective on the future of drug development. The authors argued that it was high time for a paradigm shift—towards a more continuous, integrated model that maximizes the use of accumulated knowledge through the entire development program. Central tools for implementing this model were identified: biomarkers, modeling and simulation, Bayesian methodologies, and adaptive designs—all within the arsenal of a modern biostatistician! The novel model distinguished the exploratory phase (from target discovery and validation to clinical proof-of-concept, which is essentially the scope of translational medicine), and the confirmatory phase during which modern development tools are applied to larger-scale studies to identify optimal drug dose/regimens in the target patient populations that would lead to successful phase III trials and, eventually, regulatory approval.

This new philosophy of drug development has been quickly adopted by many pharma companies, and the model continues evolving, accumulating scientific and technological advances. One essential component in this model is the quantitative decision-making support. At the pre-specified points in development, decisions should be made as to the next developmental steps; e.g. to continue or stop the development (cf. Figure 1.3). For instance, two important clinical development milestones include *clinical proof-of-principle* (demonstrating target engagement based on relevant pharmacodynamic biomarkers) and *clinical proof-of-concept* (demonstrating beneficial effect on clinical outcome or its surrogate in the identified patient population). The key translational medicine goal is to maximize success in achieving clinical PoC (in a great number of drug candidates) effectively and efficiently.

To ensure high quality decisions, well-thought out decision rules should be in place. The biostatistician (with input from the clinical trial team) is responsible for quantifying, calibrating, and formulating go/no-go decision criteria. For example, consider a PoC clinical trial comparing an experimental drug versus placebo with respect to some primary outcome. Based on the PoC readout, a strategic decision is to be made. It may be desirable to have something better that a binary (Yes/No) decision whether the trial outcome is positive or not. One can define a dual criterion for PoC as follows [39]. Let the two conditions be: C_1 =Statistical significance (say, P-value < 0.1), and C_2 =Clinical relevance (say, estimated mean difference (drug vs. placebo) exceeds a certain clinical threshold). Once the trial data become available, the conditions are evaluated and the following decisions can be made: 1) GO, if both C_1 and C_2 are met; 2) NO-GO, if neither C_1 nor C_2 is met; and 3) INDETERMINATE, if only one of the criteria is met. In the latter case, the implication may be that "data are too variable" (C_1 is not met but C_2 is met), in which case one may consider increasing the sample size; or "results are statistically significant but of no clinical relevance" (C_1 is met but C_2 is not), in which case the investigators may vote for the NO-GO decision. This approach to PoC can be much more appropriate than the good old hypothesis testing approach that declares the clinical trial a success if P-value is less than a threshold and a failure otherwise.

We expect continuing interest and investment into the area of quantitative decision-making. Use of statistical and pharmacometric modeling and simulation to evaluate the probability of success of a given development program at different stages, development of novel frameworks for strategic decision-making, emergence of new statistical software tools to enable efficient implementation of these ideas, and establishment of industry best practices are anticipated in the near future. For further reading on this subject, refer to an excellent recent book by Chuang-Stein and Kirby [23].

4.5.7 Digital development

The decade of the 2010's has been characterized by significant advances in information science and technology, and the use of digital technology tools has increased tremendously in various areas of our life. *Digital medicine* is emerging as a new discipline, with its own hot areas of research. The penetration of digital technologies into pharmaceutical R&D is also evident—companies are trying to find innovative ways to streamline their drug development processes by leveraging the power of smartphones, wearable devices, cloud-based platforms for storing data, etc. Business models in which a big pharma company is partnering with an IT (startup) company become increasingly common. The outcomes of such partnerships can be truly creative products, such as a drug–digital combination Abilify MyCite, where an ingestible sensor is embedded into tablets of a psychiatric drug Ability to track compliance to medication. This "digital revolution" provides excellent opportunities for statisticians in the pharma industry, and we outline just a few of them.

The value of digital clinical research starts with digital platforms to streamline clinical trial data management. Similar to the merit of transitioning from paper-based CRF to eCRF, going digital on a larger scale can automate recruitment and retention of patients, and remote data collection can enable site-less clinical trials (which can save lots of time and money!) On another note, electronic patient-reported outcome (ePRO) systems are now adopted by many clinical investigators. ePROs can potentially improve quality of data and have better cost-efficiency compared to the paper-based ones. As smartphones became ubiquitous, mobile ePROs can become standard very soon. Of course, these complex digital systems require rigorous testing before they start working stably and can be launched full-scale. Does this sound like an opportunity for engagement of the statistician (statistical programmer)?

Another innovation effort is development of *digital biomarkers* and *digital endpoints*. One example of an unmet need for such tools is the area of central nervous system (CNS) disorders. The clinical endpoints in CNS are paper-and-pen questionnaires which may be highly variable and subject to rater bias. Development of new digital screening/diagnostic tools for patient identification and inclusion in clinical trials, digital biomarkers that can capture disease progression, and digital endpoints that can measure treatment effect is highly desirable to improve quality of clinical trials. One good example of an established breakthrough technology in CNS is magnetic resonance imaging (MRI) in multiple sclerosis. The reduction in MRI lesion counts has been established to be a very good surrogate for the clinical outcome in relapsing-remitting multiple sclerosis (RRMS), and it is now widely used as a primary endpoint in phase II RRMS studies. Statistical challenges and opportunities for digital biomarker/endpoint development are numerous—massive amounts of data need to be converted into some meaningful measures that could be analyzed statistically. Validation is required. Currently there is a lack of standards and regulations regarding what constitutes a validated digital outcome and whether it can be used as a primary endpoint. This does sound like an opportunity!

Development of *digital therapeutics* is definitely a hallmark of 21st century clinical research. The idea is to utilize digital technologies for treatment of medical conditions. The breadth of different approaches to digital therapies is remarkable—mobile apps to deliver cognitive behavioral therapies, gaming consoles for post-discharge rehabilitation, fully integrated closed-loop systems to automatically monitor blood sugar and adjust administration of insulin in people with type I diabetes, etc. An important aspect is personalization of the treatment strategy. For example, take a smartphone app that provides a cognitive behavioral therapy. The patient starts interacting with the app, and the treatment strategy is generated dynamically, according to carefully defined algorithms, applying machine learning techniques and artificial intelligence (AI) to the observed patient's data. The idea is to facilitate implementation of the principle of personalized medicine. Of course, there are numerous challenges in development of digital therapeutics to be addressed by statisticians. How do we design randomized clinical trials to provide reliable evidence of safety and effectiveness of the digital interventions? What should be the control group?

What should be the primary endpoint(s)? What is the regulatory pathway to approval of the digital therapeutics? All these are to be addressed in the near future. We are not alone in this journey—refer to the FDA's digital innovation plan—https://www.fda.gov/medicaldevices/digitalhealth/.

4.6 Summary

In this chapter we have provided some personal opinions and strategies that may be helpful for a biostatistician working in the pharma industry. While trying to be pragmatic, we cannot avoid mentioning some extra "secret weapons".

One important element is constant learning. The field is advancing rapidly, and it is important to be up-to-date with progress in biostatistics. For this purpose, we may want to keep track of publications in major biostatistics journals (in particular, special journal issues devoted to hot topics such as adaptive designs, handling of missing data, multiplicity, personalized medicine, real world evidence, etc). On a broader scale, book series on various topics in biostatistics are increasingly common (check the websites of the major book publishers—there will be plenty). The constant learning paradigm assumes systematic categorization of the information; e.g. by topic—adaptive designs, handling of missing data, etc. There are many additional valuable sources of information—podcasts, meetups, workshops, conferences—they can be useful to establish new connections and networks.

A very powerful weapon is independent research. It is an investment that pays off only in the long-term, but the dividends can be astounding! It may be disappointing at initial stages, and it certainly requires courage, perseverance and understanding of "why I need it". The motivation can vary, but the ultimate goal is to improve problem solving skills, technical writing skills, and gain external visibility through scientific publications, presentations, short-courses, etc. The most pragmatic person would probably focus on development of some method/technology that can be patented and generate revenue in the long-run.

Finally, many established leaders in the pharma industry participate in global industry-wide efforts, such as PhRMA working groups on hot topics in biostatistics, and hold adjunct professor positions in academia. These are certainly "best of the best", but it is always good to keep this in perspective.

Introduction to Chapters 5, 6, and 7

The next three chapters represent research and development of real life questions and contain all vital parts of multidimensional problem solving (cf. Figure 1.1). A more succinct version of this process can be described as follows:

1. Recognition of problem existence.

2. Study of available methods and tools, with consequent realization that they are insufficient.

3. Clear formulation and formalization of the main task to be solved.

4. Evaluation of a solution itself, together with "the road to success (or not)" from the perspective of a "true" problem solver, along with the following:

   - Assessment of created by-products as potentially new tools, skills and methods, and especially *approaches*.

   - Improvement in understanding of neighboring fields.

   - Obtaining intellectual satisfaction and small enhancement in a philosophical system of beliefs (in extreme cases it could be the only practical results).

5. Generalization of all achievements and evaluation of potential applications in the real world.

An additional, frequently missed property of any real-world problem solving is an enormous uncertainty in the *practicality* of the results (regardless of the quality of the solution).

For instance, the outcome scoring system discussed in Chapter 5 emerged as a solution to an intermediate open sub-problem of an attempt to create benefit/risk assessment for a planned clinical trial. It later became a much more general task of safety outcome evaluation of the completed clinical trial. Thereafter, it became a topic for an FDA workshop, led to the creation of a website designed for future development and validation, went dormant for several years, and is currently considered to be actually applied in prospective pivotal trials.

The resurrection of a failed clinical program (Chapter 6) is still ongoing. According to the current development, the drug that was mislabeled from the very beginning has now a good chance to be approved with proper indication within the next 3–5 years. The main by-products include better understanding of safety/efficacy, dose–response, the nature of major mistakes that are frequently made during drug development, and continuous joint work of the two authors (AP and LBP) for the last 10 years.

The planning of open-ended projects (Chapter 7) has mostly philosophical value. The described results may help one to develop a realistic approach to planning and to lower one's expectations as far as the *honest* assessment of an open-ended task is concerned. While the method gives an opportunity to "map out" some potential road blocks, it cannot provide the accuracy that business administrators are looking for and most of the professional planners claim. In other words, since realistic planning is very useful but has no place in practice, we highly recommend using it routinely but describing the results very selectively.

Speaking of practicality, it is always a personal decision whether or not to get engaged in complex problem solving. The only *guaranteed* practical reward is honing of your intellectual and professional skills. Other important results and solutions obtained in due course (albeit they are the true reason for one's engagement) may carry little reward in practice; however different kinds of negative consequences may arise just because one is going against the current of the established practices. For the two authors (AP and LBP) who have worked together on these problems for a decade, personal rewards were mostly limited to intellectual satisfaction and significant improvement of their professional skills. (Frankly speaking, in the long run, the latter did translate into a significant material value.) Of course, during work we were saying to ourselves: "Since there are no tools available to solve our problem, we, as honest researchers, have no other choice than developing new ones." However, this is a typical white lie. Choices always exist; e.g. one may ignore the fact that a traditional tool is inappropriate for the problem but use it anyway; one may reformulate the problem such that it can be solved using the familiar tools (which is a common practice of high-end consultants); or one may even elect to completely ignore the problem's existence and just pass it by (which is, perhaps, the most frequent choice if we add cases of "honest non-seeing it"). Yet, the most important reason behind our decision (and part of our philosophy) is that problem solving does make life interesting and we enjoy doing it!

Before delving into Chapters 5, 6, and 7, we would like to make two important remarks. First, we would like to apologize outright that we have not included any references to the relevant literature in Chapters 5, 6, and 7. This was done for a reason: the references include hundreds of books and research papers, and any meaningful systematic organization of them is beyond the scope of the current book.

Our second remark is on the team work in problem solving. Contrary to a widespread opinion, complex problem solving often requires "lone wolf" efforts. Successful team work in projects that involve *very* complex problem solving may require a strong leader who does 95–100% of the thinking and delegates implementation aspects (including mining of important pieces of information) to other team members. This leader is not necessarily an official administrative project lead; in fact, freeing the thought leader from minutia has its own advantages. A collaboration of equals (AP and LBP) in the work that will be described momentarily is rather an exception (due to a complimentary nature of the professional skills and intellectual preferences). In short, complex mathematical modeling and data analysis were done by one author (AP), whereas implementation of the created models (which was much harder and required deep knowledge and understanding of drug development) was done by the other (LBP). For presentation of technical results, one author cared for the ideology, while the other one translated it from "mathematical Russian that uses English words" into "plain and well-understood by everyone business English". Furthermore, if we replace the words "business English" with the "biostatistician's English" (done by OS), we could say that the entire presentation has been revised and transformed at the next iteration into the book-length format.

Last, but not least, we would personally recommend using "95% complete" as a criterion for judgement of the project completion. It is actually a compromise between the two authors' personal preferences: one loses interest as soon as the problem is solved (in DIY terms, something that can serve the intended purpose, which is usually after $\sim 80\%$ of the project is done) and wishes to skip the boring finishing part, whereas the other one believes that there is always room for improvement and the work should be stopped only when time is up.

5

Development of New Validated Scoring Systems

"If you happen to be one of the fretful minority who can do creative work, never force an idea; you'll abort it if you do. Be patient and you'll give birth to it when the time is ripe. Learn to wait."

Robert A. Heinlein "The Notebooks of Lazarus Long"

5.1 Introduction

Let us start with the statement of the obvious. In practice, while acknowledging the act of measurement, we usually do not ask ourselves the question "What are we measuring?" The other important question "How should we do this?" is frequently replaced by "How is this normally done?" In this case, there is actually no room for problem solving unless we attribute the term "problem solving" to identifying and reading some relevant literature and then applying the acquired knowledge to our particular task at hand.

The real problem solving starts when, after successful application of the acquired knowledge, we are dissatisfied with the results of our work. We would revisit our newly acquired knowledge and the quality of its application, and even though the results seem to be a bit improved, we are dissatisfied even further.

Consequently, the next step is an extensive search in hope that someone has already addressed these issues, even though we are unwilling or unable to formulate them clearly yet. We may find what we are looking for; otherwise, we have found our problem. Of course, it all starts from us being disappointed with the results and having an inner drive for improving the situation and, as we mentioned before, it can be stopped at any time without resolution.

Next, we reach the point where we need a clear understanding of what the problem is (or where the problems are located). It stands to reason that this precedes formalization of the problem, let alone any attempt to find a solution. At this instant, the questions "What are we measuring?" and "How should we do this?" necessarily come into play. In some cases (e.g. as we discuss in this section), answering these questions is the ultimate goal, which means that this is the problem we are trying to solve. Even if this is not the case, it can enormously improve our understanding of the situation.

5.2   Recognition of problem existence

Assessment of the safety outcome of a completed RCT is a backbone of the evidence-based approach in the pharmaceutical industry. Over the past 50 years or more, it has been done for every approved or failed drug. There are some standard tools that are used in the presentation of safety data. These tools are used over and over again under the assumption that they guarantee an objective assessment, which, in turn, would lead to an objective decision making by the regulatory agencies (e.g. approve; ask for additional pre-clinical studies; put on clinical hold; request for another RCT; etc.) There are other components for regulatory decision making (e.g. health economics); however in this chapter we shall focus only on the safety outcome assessment.

The question "What we are measuring?" is asked seldom—we are measuring the "safety outcome". The question "How should we do it?" is even more seldom—we are presenting data as required. There is no problem until we actually try to measure the safety outcome.

The story started when the first and the third authors tried to quantitatively access the safety outcome of a not very successful drug development program. Safety data contained multiple safety signals and the authors were trying to develop an integrated numerical evaluation that would incorporate medical expertise and could serve as a basis for an objective decision making. The official position of both the sponsor and the FDA was that the drug was effective but unsafe. Thus, for future benefit/risk assessment of newly planned RCTs in trauma settings, we needed pure numbers to compare. There were some good indicators of efficacy (described in terms of decreased mortality, derived from an extensive pre-clinical program in trauma settings), to be weighed against the "unsafe" conclusion (derived from a huge clinical program (> 20 RCTs) conducted in various non-trauma settings). The rationale for such a comparison was based on the following assumptions:

- Safety profile of the drug is objectively described by the ISS analyses of the clinical program performed in various settings.

- Projected efficacy (decrease in mortality) is objectively described by analyses of the pre-clinical program performed in the trauma settings.

- Both safety and efficacy will be observed (independently) in the planned RCTs, as described above.[1]

---

[1] This is one ubiquitous assumption for most of the benefit/risk assessments.

All previous attempts were unsuccessful to deliver a direct (numeric) comparison, thus yielding a number of presentations that listed information about observed dichotomies in different combinations and ended up with the same "incredibly powerful" conclusion: "This is a judgment call". We were extremely frustrated, and this led us to the next step.

5.3 Study of available methods and tools with consequent realization that they are insufficient

Let us revisit the background of the problem: we already have full knowledge of the safety outcome, derived from the clinical program, that is presented by standard means:

- Difference in mortality.

- Difference in SAE incidence.

- Difference in AE incidence.

In addition, we have all possible safety signals, either calculated or available for calculation:

- Difference in any SAE and in any group of SAEs (e.g. by MedDRA Preferred Term, SOC or any other combinations).

- Difference in any AE and in any group of AEs (see above).

On the other hand, we have a projected difference in mortality, derived from the relevant pre-clinical program, which we want to compare with all vast available information. (Note: The question of how this projection was derived from multiple studies deserves a separate lengthy discussion, which is not relevant to the proposed solution.)

Besides standard reporting of safety, not many tools are available.

Obviously, there would be no problem if we were to compare only the observed difference in mortality vs. the expected difference in mortality (or any two equal dichotomies). There were papers that provided fashionable terminology (number needed to treat (NNT), number needed to harm (NNH), etc.) for this simple task. Frankly, there were some attempts to generalize it to the level of multiple dichotomies, but these attempts used such strong assumptions and vague terminology, that any practical interpretation of the results with the number of dichotomies more than 3 became virtually impossible.

At this point, we already realized that the real problem is the assessment of the safety outcome, but we were not ready to formalize it. In other words, we still did not know what we are looking for. Note that while "quantitative assessment of safety" sounds great, it provides no specifics on what we want to achieve.

Before trying to formulate what we actually need, let us see what is wrong with the standard methods of the outcome evaluation or why they actually are inadequate.

The very first question is *"Why are the dichotomies used so much in representation of the results?"* This question is almost rhetorical—obviously, because they are very simple to interpret (yes/no; good/bad; present/absent; etc.) and very convenient to analyze due to their numerical representation (1 or 0). Any statistical test comparing two treatment arms in an RCT with respect to a single dichotomy provides an ultimate evaluation of which arm is better and how significant the difference is. Of course, any statistician would be happy if the primary endpoint of the RCT is represented with a single dichotomy; yet an assessment based on multiple dichotomies may create serious difficulties.

The next question is *"What is the dichotomy from the mathematical perspective?"* This question is rhetorical for anyone who is familiar with the set theory—on the population level, any dichotomy creates two classes: those who have it (whatever it is) and those who do not have it. On the patient (element of set) level, this translates into the simplest scoring 1 vs. 0. The second property makes a single dichotomy so desirable, while the first one creates all kinds of difficulties in working with multiple dichotomies.

Let us start analyzing the difficulties from the simplest case of multiple dichotomies that are commonly used in the safety outcome assessment: mortality; SAE incidence; and AE incidence.

Of course, in the *mortality* trials (where the only question is whether a patient survived or not), two latter dichotomies will be disregarded. The other cases that provide simple answers are: "mortality rates are equal and we compare the SAE incidence" or "mortality rates and SAE incidences are equal and we compare the AE incidence".

Frequently, all 3 dichotomies (mortality, SAE, AE) are in agreement; yet they may differ by the level of statistical significance. In such cases, a conclusion can be made unequivocally that one of the arms is better.

The problems arise if some of them are in disagreement. These, in fact, do arise quite often once we want to get an "integrated" assessment of safety signals (especially if we want to attach numbers to this assessment). First of all, the signals do not carry equal importance, and, if we want to combine them, we should weigh them. Suppose, we come to an agreement that excess of 1 dead patient = excess of 10 patients with an SAE = excess of 100 patients with an AE. One may think that the only remaining piece is to sum up the weighted differences; unfortunately, this does not work because difference in mortality will be counted three times (mortality, SAE and AE), and difference in SAE—twice (SAE and AE).

The simplest case of folded dichotomies (depicted in Figure 5.1) permits the easiest solution by using mutually exclusive classes: M (patents who died),

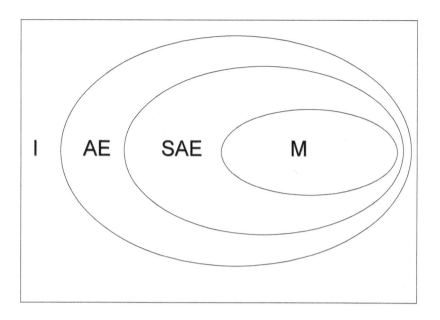

FIGURE 5.1
Study population by 3 natural dichotomies: M=mortality; SAE=Serious Adverse Event; AE=Adverse Event. Unlike random dichotomies, these 3 natural dichotomies create 3+1=4 classes: 1) M (mortality subset: M=1, SAE=1, AE=1); 2) $SAE \setminus M$ (SAE without mortality subset: M=0, SAE=1, AE=1); 3) $AE \setminus SAE$ (AE without SAE subset: M=0, SAE=0, AE=1); and 4) $I \setminus AE$ (complement from AE to the full set: M=0, SAE=0, AE=0). This provides the basis for that Mortality$\leq$SAE$\leq$AE.

$SAE \setminus M$ (survivors with any SAE) and $AE \setminus SAE$ (survivors with any AE, but without SAE). The fourth class $I \setminus AE$ (patients without any AE) is complementary and receives zero weight. Now, we can assess "how bad" each arm is by multiplying each class size (as a fraction of the entire arm) by the assigned weight. We even have a choice of unit: AE, SAE, or mortality. Finally, we can evaluate the difference between arms related to the occurrence of these natural dichotomies.

As simple as it is, to the authors' knowledge nobody ever used this approach. We can generalize this method to any number of folded dichotomies. The left side of Figure 5.2 depicts this simple case: all our signals are actually either "bad" or "good"; the vectors (safety signals) are collinear (either "+" or "−" without interaction), and the lengths of the vectors are known based on the magnitude of difference and clinical importance. Moreover, we have a strong feeling that every next signal in the hierarchy carries higher weight

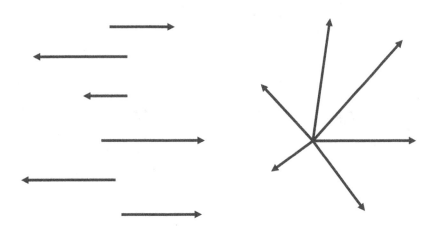

FIGURE 5.2
Left plot: a customary approach to assess safety signals which can work for
folded dichotomies. The relationship between signals within the same SOC
(or AE, or a group of AEs) creates a linear system, where all vectors (safety
signals) are collinear (either "+" or "–" without interaction). Right plot: a
more realistic relationship between safety signals. The signals from different
SOCs (or AEs, or groups of AEs) may interact (as represented by different
angles between vectors).

than the predecessor: in our case, death is worse than any SAE, which, in
turn, is worse than any AE. If we have assigned weights, then we have a fully
integrated assessment of the outcome for all mentioned folded signals. While
it is still crude, it is obviously more sensitive and objective than any single
signal used in this assessment.

Note: In theory, the longer the list of folded dichotomies is, the better—
in the extreme case, we may end up with the exact ranking of all patients
(in terms of "how bad" the patient was). The more (non-empty) classes we
create, the more accurate the ranking is. This is pure theory though—the
maximum practical improvement would be to insert 2–3 types of "durable"
or "especially severe SAE" between the mortality and the SAE dichotomies.
Frequently, this provides a reasonable incremental improvement to—but still
a far cry from—an "ultimately accurate" assessment.

Let us now move on to real-life attempts to integrate multiple safety signals, which is usually done using natural dichotomies (AE; SAE; mortality), selected groups of AEs (e.g. cardiac SAE), and "elementary" signals of interest (e.g. myocardial infarction, stroke, etc.) As we have seen for folded signals, the number of classes to weigh is no more than $n + 1$ for n signals. For non-folded dichotomies, the theoretical number of classes is no more than 2^n, and the number may go down for mutually exclusive dichotomies. Elementary dichotomies that are used most (differences in specific AE or SAE incidence) are particularly bad for the number of classes to consider. For instance, Med-DRA, has about 800 terms, which are neither folded nor mutually exclusive. This means that 20 "elementary" signals of interest will create in our population 2^{20} classes to evaluate in order to make a valid integration. Adding the other common dichotomies will increase this huge number even more. To make things worse, on the top of inequality of signals, we are losing one-directional assessment—there is no simple "bad" and "good" anymore; now our signals are depicted in the right part of Figure 5.2.

It now becomes clear that any attempt to integrate safety signals on the population level is doomed. This leaves us with the only remaining possibility—to create an assessment on the patient level.

5.4 Clear formulation and formalization of the main task to be solved

So, we are looking for a numerical assessment of the safety outcome for every patient. In order to formalize our task, we need to step back and look closely (and systematically) at all existing scoring systems. Although there is an incredible variety of scoring systems, they can be classified into several major types according to their goals, means, complexity, approach, and numerical properties.

I. Goals: assessment vs. prediction

An *assessment* scoring system answers the question *"What do we have?"* and usually involves the use of multiple dichotomies observed during some period of time, which are sub-outcomes to be integrated into a single assessment. A *prediction* scoring system answers the question *"What do we expect?"* and usually involves the use of multiple covariates (dichotomies included, but ordinal and continuous covariates are commonly used) to predict the outcome of interest.

Note: Sometimes these categories can be confused, because assessment scores also have some predictive power, and prediction scores assess (to some extent) what has happened. In other words, the results could show a high degree of positive correlation. Nonetheless, the crucial difference is in the proper selection of baseline covariates. For instance, *age* is a ubiquitous covariate in prediction scoring, but it can be a classification factor (to establish subgroups) in assessment scoring. In addition, assessment scoring evaluates change from baseline rather than the outcome itself. Furthermore, in a randomized (and thus balanced at baseline) trial, the direct assessment of the outcome is a valid substitute for assessing the change from baseline, and vice versa.

II. Means: population vs. subject

Prediction scoring is naturally subject-centered; that is, the goal is to predict the outcome for an individual. An example of prediction scoring is the ISS score utilized in the trauma setting. Assessment scoring explores either the population or individual subjects. The well-known NNH and NNT outcome scoring (even with the assumed conversion factor of 1, risk vs. benefit, it is still a scoring system) or juggling multiple dichotomous sub-outcomes represent population-oriented assessment scoring systems. By contrast, the blinded medical assessment is an example of the subject-oriented assessment scoring system.

Note: Due to the probabilistic nature of the tools used, one can expect reliable conclusions only if the population size is sufficiently large. Accuracy of the scores to characterize any individual subject (for both assessment and prediction scoring) cannot be assured with 100% certainty.

III. Complexity: number, nature and structure of covariates

A potential pool for covariate selection is limitless. It contains all kinds of dichotomies: ordinal and continuous variables can be structured in an infinite number of ways. Nevertheless, their relevance and feasibility restrict the number of covariates for all population-driven (safety outcome analysis) and calculation-based (e.g. safety outcome modeling using some regression methods) approaches. The only exception is the *blinded review* (scoring individual patients by a panel of experts, followed by a statistical analysis of the obtained scores)—it is well known that never does one have enough data for accurate and complete assessment—no wonder that blinded reviews are usually a logistic nightmare. Excluding the case of a single covariate (which does not require scoring), the scoring usually involves 2–10 covariates for finite systems and an infinite number for the blinded review.

Note: The methodology proposed by the authors actually fills this gap consistently—balancing relevance, completeness, and feasibility, while accounting for medical expertise.

IV. Approaches: simplistic; formula; model fitting; panel

Simplistic: This represents the sum or maximum (trumping) for the observed dichotomies, and the selection of dichotomies could be derived from panel (uncommon) or from one–two medical experts (common).

Formula (calculation): Weighted sum or other simple formula (e.g. sum of squares in ISS, trauma scoring). Weights are commonly assigned by a panel, but selection of covariates (cf. *Model fitting*) and especially the formula properties are very unlikely to be provided by a panel.

Model fitting: A major tool for prediction scoring is the *model*. Clinical expertise may be used only for selection of the preliminary pool of covariates, but this is mostly a theoretical possibility. In practice, this pool is usually defined by the availability of data, and final selection is commonly based on demonstrated significance (e.g. from stepwise algorithms).

Panel: Taken alone, this is a pure medical assessment. Blinded panel review of every subject, despite many shortcomings, is recognized as the most straightforward and reliable method for outcome assessment. It is almost never used for prediction scoring in practice, but it actually has higher predictive power than any fitted model.

Note: The probabilistic approach is built into any panel evaluation; therefore, the most reliable results are observed for sufficiently large populations.

V. Numerical Properties

The scale used for any scoring system is almost irrelevant, and any system can be normalized into 0–1 scale. Sensitivity is incomparably better for continuous scales, but reliability of limited (3–6 ranks) ordinal (categorical) scales is generally higher. The natural requirements for symmetry and cumulativeness are usually met "by default" in all scoring systems. Asymptotic requirements, as a rule, are met for subject-driven scoring, and in the majority of cases it is broken for population assessment (e.g. summing up the incidences of several dichotomies can easily exceed 1 (or 100%).

The major problem of population-driven scoring is the *inability to account for interceptions of the observed dichotomies*.

The difference between continuous and ordinal scales goes beyond the range of possible values. Actually, there is a possibility that a scale with just 5–6 values can be continuous by nature, whereas a 100-value scale can, in fact, be ordinal. The real test of whether a scale is continuous is the question of whether the distance between two values has a meaning or not. If a scale simply establishes the order, but does not assess the magnitude of the difference, then such a scale (even with a very large number of possible values) remains ordinal. This means that categorical tests are applicable to this scale, but continuous tests are not. Mathematically, in order for a scale to be continuous, it should possess some linearity—the same distance between values at any interval of scale should yield approximately the same difference.

The best chance to achieve such linearity is to have a scoring system based on recurrent formulas that are linear at the lower end of the scale. For instance, the formula $x + y - xy$ has a good chance to create a continuous scale, whereas $x^2 + y^2 - xy$ does not.

Note: Regarding multiple applications of scoring systems, it may seem that an assessment of survival time adjusted for the quality of life, health, etc. is not the subject of the categorization scheme described above. However, this type of evaluation is precisely based on the proposed approach: the 0–1 scale (just reversed in order to create an easily interpretable AUC) is applied to multiple time intervals and an adjusted survival time could be calculated.

According to the above classification scheme, our task could be formalized as finding a scoring system that has the following properties:

I. Goals: Outcome *assessment.*

II. Means: *Subject*-driven (but should be used for populations).

III. Complexity: Feasible and flexible with a range of complexity driven by the relevance and completeness requirements.

IV. Approach: Exploring the *formula–panel* approach when the panel evaluates elementary events (covariates).

V. Numerical properties: Using formulas with pre-defined properties (symmetrical, cumulative and asymptotic) with the preferred scale 0–1.

5.5 A solution itself

In §5.4 above, the selections for items III, IV and V are straightforward, whereas the selections for items I and II require more care. Let us see how it all worked out (read on and be intrigued!) The very first thought after looking at I and II above is: *"Since we have to incorporate clinical expertise, what is wrong with the blinded review?"* The counter-arguments are:

- It is extremely expensive. Consider a study with $n = 800$ patients, with a panel of at least 3 reviewers per patient, and at least 0.5 hours per reviewer per patient. This gives us ∼1,200 hours of highly paid ($300–$400 per hour) personnel, to begin with. If one adds all preparatory work, database design, data collection and proper blinding, the price tag can easily be in the range of $1–2M.

- It is extremely time-consuming.

- It is biased—it is impossible to have a panel of the same 3 reviewers for every patient, and (irrespective of how much time and money is spent on developing a common strategy for the review), any reviewer will assess the same patient differently. In addition, having 3 reviewers is not enough to address random or systematic deviations in individual scores, and more than 3 reviewers will increase both the cost and the logistical difficulties.

- One may have to choose between proper blinding and withholding of important (and sometimes critical) information. For example, in a trial comparing the HBOC against the pRBC, not only should the plasma Hb (which directly reflects the volume of the HBOC on circulation) be removed to assure proper blinding, but also any lab report that contains both the HCT and the THb should be "doctored"—one of the two most important hematology markers must be removed (as THb[G/dL]>HCT[%]/3 implies significant volume of HBOC in circulation).

- The commonly used scale for safety assessment is 0–5 (0=no problem; 1=mild; 2=medium; 3=severe; 4=life-threatening; and 5=death). This is a typical ordinal scale without linearity, which is not very sensitive for detection of treatment difference (albeit it is better than a simple dichotomy).

A good thing about blinded review is the fact that the panel does not evaluate populations (arms), but rather each individual subject. A bad thing is that a lot of information should be evaluated in order to get a simple score. A proper question is *"Where are the time and efforts spent most?"* Of course, they should be spent on on getting from the full patient dossier to the final score. It is understood that 90–95% of information is ignored, but 100% of information should be processed. In reality, every reviewer develops their own algorithm that permits finding quickly the relevant information and assigning the score based on it. Let us not forget that the original purpose for the team of reviewers is to have a single algorithm in place (or, realistically, very similar ones) that would produce essentially the same score.

The "ideal" solution looks almost unavoidably evident:

1. Select limited, but complete and sufficient amount of information.

2. Break it into a reasonable number of elements.

3. Have the panel to assign a score for every element and create the final score based on the panel results.

4. Develop an algorithm to combine individual scores.

Note that this algorithm should be objective and recurrent, based on combining any two scores into one using common sense requirements for such combinations.

We leave to the reader a comparison of this solution with our proposed task formalization in §5.4.

While on the ideological level the problem has been solved, we must assure that every step actually can be performed and go through a lot of details and auxiliary problems. What we have now is the foundation for the future solution—a plan. At this point we realize that we attempt to create a tool (even methodology) that permits an objective assessment of the results for both planned and already conducted trials.

The formal description of the solution which we shall describe momentarily is an enhanced version of what was done and was written up several years later when we fully understood that we had at hand a new methodology with a broad spectrum of possible applications.

The desired properties of this tool are:

- Creation of a comparable objective score for every patient.

- Incorporation of clinical evaluation without individual bias.

- *Flexibility* – the ability to tailor the scoring system to a specific situation.

- Balance between *sensitivity* (preciseness) and *reliability* (relevance).

- *Transparency* (traceability) – the ability to demonstrate integrity of the tool.

- *Efficiency* – both time- and money-wise.

The main idea of the proposed solution is to change the target of clinical evaluation. In the process of blinded review, the target of clinical evaluation is the entire patient safety profile. In our presented tool, the target of the clinical evaluation is an AE or a group of AEs.

In general, the mechanism is to use a panel of experts to assign a score to each AE according to a specific trial risk tolerance and then programmatically calculate the cumulative score for each patient. This mechanism eliminates all questions related to bias introduced by clinical evaluation on the individual level, since all patients would receive scores developed for a particular study using the same rules by the same panel of experts.

Of course, we have two "small" problems to solve:

- *What is the appropriate level of scoring?* Obviously, in high mortality trials, clinical evaluation will be very little affected by non-serious AEs; thus it does not make any sense to score hundreds or even thousands of events with minimal clinical relevance.

- *How should the cumulative score be calculated?* It cannot be a simple sum and it potentially depends on the answer to the previous question.

In order to answer the first question, we added an extra (first) step to our tool. Before conducting a survey with the entire panel of experts, we had a small "round table" discussion with two experts (many thanks to Doctors Gerson Greenberg and Leo Kaplan!) from the planned panel. The goal was to define the balance between required preciseness (sensitivity) and relevance (reliability). We learned during this preliminary conversation that trauma surgeons were interested (besides mortality) in the incidence of SAE for every SOC, and, to some extent, in the incidence of any AE. In addition, we realized that the risk tolerance highly depends on the mortality level. These findings were used while creating specific scoring systems for various levels of mortality and actually enforced the design of the questionnaire. (Again, many thanks to Cdr. Dan Freilich, who was able to line up for us 10 more medical experts and get the results of the panel evaluation within several days.)

To answer the second question, we needed a clear formulation of requirements for a meaningful scoring system.

1. *Choosing the scale.* The choice was to use a natural and easily interpretable mortality equivalent, which called for the 0–1 scale.

2. *Choosing the recursive formula for adding up two or more scores.* There were some requirements that had to be met in order to choose a viable formula. Generally, all of them were determined by common sense:

    a) All scores are positive numbers from 0 to 1.

    b) The result should depend neither on the order of scores nor on the order of summation for any set of scores. Mathematically this means that a function of two variables, $F(x, y)$, that calculates the cumulative score for any two given scores should be fully symmetrical: $F(x, y) = F(y, x)$ and even more $F(F(x, y), z) = F(x, F(y, z)) = F(F(x, z), y)$.

    c) The death (score=1) should be the absolute maximum that could be reached only if one of scores in the sum is 1. Mathematically this can be expressed as $F(x, 1) = 1$, and $x < 1$ and $y < 1$ implies $F(x, y) < 1$, and the function asymptotically approaches absolute maximum value of 1.

    d) Adding a score of 0 should not affect any score: $F(x, 0) = x$.

    e) The function should actually yield a score that is higher than any of the scores summed: $F(x, y) \geq x$ and $F(x, y) \geq y$. The only possibility for the "=" is when one of the scores is equal to 0 or 1.

    f) While the asymptotic requirement means that for scores that are close to 1, adding an extra score increases the cumulative score very little (e.g. $0.95 < F(0.95, 0.5) < 1$), for small scores it is preferable to have $F(x, y) \approx x + y$.

Mathematics provides an infinite number of functions that meet the above requirements. We restricted our choice to the two most common functions that have been proven to do the trick in very similar settings: *relativistic* (summing velocities in relativity theory, when they are expressed as a fraction of speed of light—the mighty absolute that cannot be surpassed): $F_R(x, y) = (x + y)/(1 + xy)$ and *probabilistic* (probability of event $A \bigcup B$, where $\Pr(A) = x$, $\Pr(B) = y$, and $\Pr(A \bigcap B) = xy$): $F_P(x, y) = x + y - xy$.

It can be shown that $\max(x, y) \leq F_P(x, y) \leq F_R(x, y) \leq 1$. In our first iteration, we chose the relativistic formula since it penalizes multiple AEs a little more. Frankly, the second formula is simpler and is probably more appropriate for measuring distance between the scores. More detailed discussion of the properties of these two formulas is provided in Appendix A.

3. *Choosing the system for AE coding that would provide, at the same time, completeness and avoidance of multiple counts.* In general, any complete two-level coding system should do the work, since it provides the ability of accounting AE (SAE) on the level of a separate term or on the level of the related body system. We decided to use MedDRA as it is a well established industry standard. Theoretically, it is even possible to combine or separate terms from any available coding system into new groups according to recommendations of the "round table".

   In our case, after consulting with experts, we decided to separate for the scoring purposes the "durable SAE" from another SAE of the same SOC. The term "durable" was used for an SAE (e.g. myocardial infarction, stroke, renal failure) that could not be resolved completely without any consequences. This was important because the major assumption for all other SAEs was that they were ultimately resolved, which means that the patient either died or had no long-term consequences. Actually, this was the reason why we did not separate cardiac arrest from the entire Cardiac SOC.

   Finally, after our "round table", we decided to use a higher score for the entire SOC with presented durable SAE. In other words, the presence of "durable SAE" trumped the entire SOC. Technically speaking, the issue that forced us to separate "durable SAE" was the fact that it was almost impossible for the panel members to assess the relative frequency of such SAEs.

   In order to avoid multiple counting of exactly the same or similar events, it looks reasonable to not count more than once any AE. However, the last word in question, whether we should count the same (similar) AEs more than once, belongs to the the experts.

4. *Assigning the final score to an elementary covariate (AE or group of AE) based on results of the panel.* Usual solutions for panels is to use either median or mean either for all collected scores or after removing the smallest and the biggest scores. Essentially, it does not matter—we decided to use mean after removing extremes. A small detail: we could not ask the panel to use the 0–1 scale because not everyone feels comfortable with decimals. Instead, we proposed using the 0–100 scale that was later converted into 0–1.

5. *Building a hierarchical structure for application of the formula.* Note that the presence of multiple levels of scoring may create additional problems while calculating cumulative scores. For example, in our trauma application, we accounted for AE incidence only for patients without SAE. In

addition, we had a 2-step scoring for 3 SOCs that contained "durable SAEs". The situation may become more complicated if we are advised to evaluate, for example, separate SAE in some SOCs and score the entire SOC for others. The methodology permits various approaches to this technical problem:

a) We may decide that SOCs that were not divided are equally important to selected terms and proceed exactly as we did for the SOC evaluations.

b) It is possible to score entire divided SOCs separately in a two-step process, e.g. using the absolute maximum assigned to the SOC.

c) Etc., etc.

The major point is that almost any requirement from the experts can be accommodated relatively easy in the survey, and later into calculation of cumulative scores.

6. *Writing a program to combine scores for every patient.* This is, perhaps, the easiest part of the entire solution. As a starting point, we used the standard MedDRA-coded AE database and the dataset that was created from the survey results. Applying the designed earlier algorithm was simple enough. In fact, since we had 3 different scorings for various levels of mortality and wanted to test both median and mean final scores, we ended up with 6 scores for every patient in the final database.

5.6 Are we finished? Not in the regulatory setting!

A major prerequisite for future acceptance of your work is complete and detailed documentation of the entire project. Documentation for the survey and the SAS programs that were used to implement the algorithms are standard and deserve only brief mention here. The biggest problem was a comprehensive description of the project rationale as part of the final report. The latter consists of just one word—*validation.* A newly proposed methodology constitutes a conundrum: you can use it if you can demonstrate that it has been successfully implemented before. In reality, depending on your personal status and the friendliness of your reviewers it can be anywhere between "Great idea, go ahead, your application will be the validation of new methodology!" and "No, you cannot use a non-validated methodology, ever!"

Intermission: In the authors' humble opinion, there are three stages of rejections of attempts to promote a valid/good novel approach:

• This is wrong! (I am not buying this!)

- This is correct, but it is not needed/non-practical.

- This is trivia; it was implemented so many times that there is no novelty.

As any good joke, this is only partially a joke. Regretfully, any "lone wolf" who attempts to promote new ideas—without securing in advance support from important stakeholders—will likely face it. The most regrettable feature (and at the same time the most promising feature for science/industry) is #3. Since the process may take decades, the #3 rejection could be the only truthful statement of the three.

In theory, validation means demonstration that a proposed tool actually works and produces scientifically sound results, and when tested in parallel with other tools, it achieves similar results. Showing better results may be a slippery slope, because "better" might be interpreted as "not similar".

In order to test the obtained scores, we used an extensively analyzed mid-size (700 patients) pivotal trial with conducted blinded review. The safety conclusions from this trial were well established: overall arm A was better than arm B with marginal statistical significance; two paired subgroup analyses demonstrated one pair of subgroups (A_1 and B_1, near 60% of patients) was incomparably better than the other (A_2 and B_2, about 40% of the trial). In the bigger pair A_1 and B_1, statistical equipoise was generally maintained, but in the complementary pair A_2 and B_2, despite a significantly smaller size, arm A was much better (practically all tests showed statistical significance). The nature of the results provided a possibility to test both reliability and sensitivity, compare the results with the blinded review, and analyze trends caused by applications of multiple scoring. In addition, testing data permitted checking the "worst case" scenario for application, because the observed mortality was at the level of 2–3% and required much higher sensitivity than we originally incorporated.

The results of this testing were very promising. Ranking tests demonstrated full agreement with the blinded review and sensitivity was almost independent from the applied scores, which indicates the exceptional ability of the scoring system to assign proper ranks to every subject. Continuous tests demonstrated increased sensitivity while moving from 90% to 30% mortality. The 30% score was enough to reproduce all results from the blinded review, which promises a continuous nature of the developed scores (something that blinded review is missing).

Later development: Five years later, one of the authors gained access to the database of a large RCT (700+ patients), which was conducted approximately at the time when the described safety outcome scoring system (OSS) was created. The trial had about 15% mortality, was already evaluated by the FDA, and, of course, had the MedDRA-coded AE database. While mortality in this trial was less than the numbers that were used in the first run of the OSS, actual expected mortality for this trial was exactly 30%. In other words, it was an ideal opportunity to test the created scoring algorithm. In short, all conclusions that were made during evaluation of this RCT by the FDA, were reproduced after applying the described OSS. The sensitivity of analyses in any subgroup was superior in comparison with any of the standard dichotomy used in the analyses. While the original conclusions were based on strong[2] trends in mortality and morbidity (AE/SAE) differences, the 30% scoring from the OSS showed a statistical significance of differences for both ranking (Wilcoxon-test) and continuous (t-test) approaches.

5.7 Assessment of created by-products as potentially new tools, skills and methods

First of all, the entire methodology (OSS) is a by-product of the original task of benefit/risk assessment. The final system includes several integrated methods: use of medical expertise to design the survey, conduct of the survey, the algorithm for combining individual scores, the accompanying software, and extensive testing/validation of the obtained scores.

The other by-product was a working classification of existing scoring systems, which includes minimal requirements to their viability (which, by the way, are not always in place for existing scoring systems). Additionally, a class of functions that could be used for combining elementary scores was described.

The preliminary results warrant at least non-inferiority of the new approach compared to the existing gold standard (blinded review), let alone expanding perspectives of usage (e.g. situations where application of results from the blinded review is impossible).

For potential advantages of the OSS over the blinded review, see the answers to the question *"What is wrong with the blinded review?"* at the beginning of §5.5. One additional advantage is the possibility to create a customized OSS for a particular setting at any time (before or after RCT) without breaking the prospective nature of the experiment and without introducing any bias.

5.8 Generalization of all achievements and evaluation of potential applications in the real world

After getting such promising results and summarizing all experience and knowledge related to scoring, the authors realized that they found an ulti-

---

[2]They were rather convincing when combined, and almost none of them standing alone was statistically significant.

FIGURE 5.3
Outcome scoring system (OSS) logo

mate solution that can utilize best properties of scoring and, at the same time, incorporate medical opinion [84]; cf. Figure 5.3. In addition, there is a definite opportunity to create a system of rules that permits generation of inexpensive, similarly designed scoring systems that could be tailored to any situation. Created scoring systems are reusable for clinical settings that they were designed for.

There is an obvious possibility to develop a new industry standard based on extensive study of the proposed approach. In such a case, there are big opportunities to streamline the safety outcome assessment and standardize it. It is possible over time to create a *library of systems* tailored for any clinical setting. It is possible to compare different drugs/therapies that have never been tested head-to-head under the same RCT conditions using observed differences against a third common drug/therapy—normalization of conditions for different RCT does not seem to be a hard task to solve. Of course, such theoretical comparison cannot serve as a basis for approval, but it is a very valuable benefit/risk assessment while planning new RCTs.

While the proposed tool does not replace regulatory decision making, the standardized approach that integrates various safety signals could be extremely helpful in decision making. For example, there is a possibility of using the difference between arms in scoring as a secondary or even a primary safety endpoint. As the latest development, this approach is under consideration for two planned RCTs.

6

Resurrecting a Failed Clinical Program

"Certainly the game is rigged. Don't let that stop you; if you don't bet, you can't win."

Robert A. Heinlein "The Notebooks of Lazarus Long"

It is almost impossible to analyze all the reasons why clinical development programs fail—the reasons are numerous and they may work together in different combinations. The authors will discuss a particular clinical program that failed due to the absence of a real PK model, exaggerated expectations, a wrong label, and misinterpretation of the RCT results. Then we will deliberate on the difficulties of a) uncovering mistakes; b) developing theory; and c) convincing top management of the company that mistakes had been made and there was a solution, which required label reformulation—all in the setting of a one-drug company. Finally, the entire field of similar products prone to the same mistakes will be examined.

A personal disclaimer: the salvation of a failed clinical program is an enormous and extremely complex enterprise that requires coordinated efforts of many dedicated people. The authors neither imply that the described case was done solely by them, nor do they attempt describing this process in full. Our modest goal is to demonstrate how the problem solving approach can help to identify and resolve (albeit from a purely theoretical standpoint) key problems that (in the authors' opinion) led to failure, and to generalize the encountered mistakes in order to avoid some of them in the future.

6.1 Preamble: what we are dealing with

First, let us describe the magnitude of the problem. The clinical program of developing a blood substitute had been almost two decades in development before it actually collapsed: the BLA submission to the FDA ended up with a permanent clinical hold that was never lifted in 6 years, during which the company tried to avoid bankruptcy.

Formally, everything was good before the submission. The pre-clinical program included near 200 animal studies that demonstrated both efficacy and safety; the clinical program consisted of 21 studies starting from phase I and including two pivotal phase III studies. Based on a smaller pivotal RCT ($n = 160$ patients), which demonstrated border-line competitiveness of the new drug vs. the RBC in treatment of severe anemia in general surgery settings, the drug was approved/registered in South Africa. Subsequently, a much larger pivotal RCT ($n = 688$ patients) was completed. The company was expanding, multiple CROs were involved in preparation of the BLA; everyone was busy with technical details (including two authors (LBP and AP) who worked independently on pharmacology and statistical analyses, respectively); the stock went higher and higher...

Of course, there were some disturbing signs: there was already a 3rd CRO that finished data cleaning and statistical analyses of the second pivotal trial; the statistical department in the company was created less than a year before the planned date of the BLA submission; previous data analyses from 20 clinical studies and study reports were done by multiple CROs; the creation of the integrated summary of safety (ISS) was extremely difficult, and non-compatible study designs and data collections from different studies did not make things better. The most important problem was in the pivotal study results: while the efficacy endpoints were met, the safety was (according to the senior management) only border-line adequate.

Despite the fact that some personnel involved in data analysis questioned whether data quality coupled with the not so good safety outcome warranted BLA submission in a hurry (which, frankly speaking, would still be 2–3 times longer than industry standards for big pharma, and unsurmountable for a small pharma company with limited resources), the official company viewpoint was:

1. The quality of data is acceptable. (Again, the first author many years later worked on submissions for a big pharma company—as a statistician—and can vouch that submissions with data of similar or even lower quality are not uncommon.)

2. Even though the results might not support unequivocally the desired label, we should be able to negotiate a restricted label during evaluation of the submitted BLA, and the more we ask from the beginning, the better our chances are to get something valuable as a tradeoff. (While this may look to be a viable strategy, it can backfire—the submission may not be taken seriously at all and might be rejected outright.)

3. Importantly, there was pressure from investors/stockholders, who had been notified long in advance that, in essence, the results of the phase III pivotal trial were great; approval of the drug was just a formality; and the date for the BLA submission was set.

Paradoxically, the real work on evaluation of the results for the entire clinical program, including reconciliation of collected data and data analyses, review of basic science behind the scene and decisions on the path forward started only after a crashing FDA letter that contained hundreds (!) of questions that must have been addressed before continuation of development. These questions essentially forced re-evaluation of the entire BLA at first and the entire clinical development program next.

For two lunatics (AP and LBP) who decided to dig out the truth, it became a many years long, exciting and exhausting, frustrating and satisfying affair that re-shaped completely their understanding of the fields they were working in, and, in a more general sense, led to re-assessment of the place that evidence-based science takes in drug development.

We shall try to keep two lines ("problems solved" and "lessons learned") separate just because this is the case in real life. Problem solving usually leads to better understanding, but it neither mandates nor guarantees such an achievement.

6.2 Problems solved

Along this journey, the authors had to formulate and solve (or not) a lot of different problems. Some of them were tactical and had limited value besides pure demonstration of how the methodology of problem solving works, but several of them actually called for further generalizations, with major lessons learned related to the HBOC field, the transfusion field, and even the entire drug development. For such problems, the solution itself becomes much more important than a particular sequence of events that led to it. In some sense, these solutions create a proper methodology that could be applied in similar situations (which is, strictly speaking, not problem solving anymore) or could be used as an example of what to look for when facing completely different situations.

The major lesson that we learned is on the level of common sense, but it is often rejected in a regulated environment (such as drug development):

> **Standard tools do not work at all in non-standard situations.**

The authors were lucky (or not) to work in such situations and found the hard way that the burden of proof in this case is almost insurmountable—you have to prove that none of the approved (and validated) tools are applicable, have new tools developed, validated, and then approved (which may take years, if not decades). A significant fraction of problems we faced, at least partially, originated from non-standard situations, with a wide spectrum of proposed solutions from "do not apply standard tools here" to "use the newly developed tools instead".

Two big problems that were associated with creation of new tools were covered in detail in the previous chapters—the data cleaning project (§3.7) and the creation and validation of the safety outcome scoring system (Chapter 5).

In what follows, we shall discuss mostly science-related problems while trying to cover both clinical and statistical aspects:

- Studying drugs with dosage that depends on needs (§6.2.1).

- Separation of toxicity and efficacy effects in the safety outcome misbalance (§6.2.2).

- Creation of a PK model for the transfusion field (§6.2.3).

- Mystery of the transfusion trigger (§6.2.4).

- Rise and fall of the HBOC field (§6.2.5).

We shall try to keep transfusion-related specifics to a minimum that will permit discussion of the following important aspect:

There are certain limitations of evidence-based science in clinical research.

A simple acknowledgement that limitations do exist is a good (and brave) beginning.

First revelations (in a long list of common reasons) for failed clinical programs came almost immediately as something shocking and unique for the "unlucky" company that the authors worked for; yet later they were also confirmed by decades of collaboration with different small pharma companies.

One revelation was related to data and data analysis: *Small (successful) pharma at the time of marketing application may not be in full possession of their own data and data analyses.* This seems so strange that it definitely requires explanation. A significant part of research and development in a small company is outsourced to multiple CROs or individual consultants. As a result, even for a single study, multiple CROs may be involved in the trial design, trial conduct, data collection, data cleaning and data reporting. Sometimes, even interpretation of the results along with medical writing is outsourced. There are occasions when the company has, as a final product, only the CSR, while both the original and cleaned databases still reside on the CROs' servers. The longer and the bigger the clinical development program is, the worse—there is simply not enough manpower and time (and, often times, experience) to request, obtain and reconcile all these issues properly. On top of this, getting programming packages either for data cleaning or data analyses usually meets the problem of intellectual property—in reality CROs just create a situation when the simplest questions that the company may have requires their involvement. It is also true that many big pharma companies often make a similar mistake by distributing workload between different teams and different departments without proper preservation of continuity in data collection, data cleaning and data analyses.

Another revelation was at the other side of the spectrum: *Frequently, there is no inclination to review the basic science behind the clinical program or the chosen direction of development until all other possibilities are completely exhausted.* The warning signs are simply ignored due to time constraints and outside pressure. This is obviously not caused by objective limitations—after all, it is frequently done after crushing failure and it costs much more. The big pharma can avoid this pitfall by assessing the feasibility of future development. As a negative consequence, some promising programs may be terminated without careful evaluation of whether the program could be improved or not. For the purpose of advances in science, the results are equally bad.

Before delving into discussion of the problems that we faced and solved, let us describe the clinical setting and the family of drugs that were studied. Our apologies to the readers who are already familiar with the subject.

Severe anemia 101

While anemia is defined by the WHO as THb < 13 g/dL, we are dealing with *severe anemia* that normally requires blood transfusion. The major cause of severe anemia is blood loss, either due to surgery or trauma (or both). Relatively small blood loss (up to 20–30% of blood in circulation) usually does not require transfusion of blood to replace RBC content—non-oxygen carrying solutions (volume expanders) are used instead to restore volume in circulation. This, of course, causes *hemodilution*— THb concentration drops by 20–30%, and if THb was normal at the beginning, it stays at the level of 10 g/dL or higher. This level, in most cases, is considered as safe and lost RBCs are restored by body in approximately 7–10 days. After bigger blood loss (or with lower starting THb), blood transfusion is prescribed for sequential volume restoration (and subsequent THb restoration). The question of when (at which level of THb) to start the RBC restoration instead of volume restoration is one of the most debated questions in the transfusion field. The correct answer is called the *transfusion trigger*. Obviously, the transfusion trigger depends on a patient's condition, co-morbidities and demographics. Existing guidances essentially agree that the transfusion trigger is located between 6 and 10 g/dL (depending on the patient and the settings), and a consensus (over the last 2 decades) is to lower it as much as possible to avoid overuse of blood and complications (mostly related to disease transmission).

The HBOCs field

Hemoglobin-based oxygen carriers (HBOCs) were developed in the 1990's, generally as an answer to a looming problem of compromised blood banks. The main difference compared with the blood is the free Hb not associated with the RBC, which behaves similarly to the RBC Hb in terms of oxygen delivery to tissues, and has some pros and cons compared to the RBC (e.g. the molecules that carry oxygen are much smaller and can reach occluded vessels; the logistics of storage and usage is much better—long shelf life, no need for compatibility, some could be stored at a room temperature; however, the major con is that they are all vasoactive and have short half-life in circulation). Most of the HBOCs failed—usually after pivotal phase III RCTs; some survived (including Hemopure that the authors studied); some new ones are currently in development. Initially, HBOCs were heralded as blood substitutes that are even better than blood; at some point they were ruled to be toxic; and currently they are aimed to be used when the blood is not available.

    Let us start with two objective reasons that created a non-standard situation (not necessarily unique!) for which we were unprepared: *Some of the standard tools (read: existing regulations and recommendations) did not work as intended.*

6.2.1   Studying drugs with dosage that depends on needs

The existence of such drugs is evident—in the transfusion field, all oxygen carriers (either blood or HBOCs) that are used to treat anemia are administered depending on the severity of anemia. The more anemic patient is, the more oxygen carrier he or she needs. The problem is rooted in the necessity to adjust some current regulated tools. Unfortunately, from the authors' experience, these adjustments were largely ignored by the regulators and, surprisingly, by some well-respected researchers in the transfusion field as well.

    Let us start with an obvious fallacy that can be found in some published peer-reviewed papers that argue for decrease of blood transfusion using the following logic: "*Let us compare the outcome (either mortality or morbidity) for patients who received 1, 2, 3 etc., units of blood.*" Unsurprisingly, such analysis shows that the more units of blood the study patients received, the worse was their outcome; thus, the blood transfusion causes bad outcomes! Some of these articles even mention in the "Limitations" section the real reason behind those findings: "*The more anemic patients are sicker to begin with, and the dose of blood they received reflects severity of anemia.*" Logically, there is no way to account for a patient's condition (even while using fashionable propensity scoring) that ignores the most important covariate in the analysis, and even if it is accounted for, there is nothing left to compare.

    In fact, such researchers just follow the conventional development path which calls for dose–escalation studies at early development stages and dose–response analyses in pivotal studies. Moreover, the medical reviewers with whom we interfaced tended to recommend restricted usage of the drug based on their interpretation of the dose–response analyses. We would say that we all are extremely lucky that blood transfusion has not been a subject of modern requirements for drug (or biologic) development! Judging by what we have seen, the blood would be deemed toxic and limited for use by 1, or at maximum 2, units.

Another not very well recognized consequence is the difficulty to properly design and analyze RCTs for oxygen carriers. As we already described in the data cleaning project (§3.7), there are two natural timelines for data collection and analyses: by real time (say, by days), and by dose (say, by units of drug). Both of them are important, and neither of them should be ignored at the design stage, but the integrated interpretation of the results is very challenging (and almost impossible to explain to a non-prepared reader or reviewer).

The good news is the fact that the dose received actually serves as a natural propensity score because it reflects the severity of the treated condition.

6.2.2 Separation of toxicity and efficacy effects in safety outcome misbalance

Suppose we have an RCT where the safety outcome is clear: one arm is obviously better than the other. No one cares about the reasons why this is the case if we are on the winning side; however, if we are on the losing side, the correct answer may change the fate of our drug. The conventional interpretation of the results is "Your drug is toxic, end of story." It originated from an ideal RCT—comparison versus placebo, where it is the most likely explanation of the outcome. The situation changes dramatically if a new drug loses to another drug that has already proven efficacy—this is exactly what happened with the HBOCs when they lost in competition with the RBC in head-to-head comparative RCTs in severe anemia settings. At first glance, the question "Why?" is irrelevant—if you lost, then you lost, and your drug should not replace the winner that works as intended. But what if you decide to change a label and use your drug for an indication where the winner is not available? Let us face it—if a drug is labeled as toxic, the clinical program is practically doomed. Theoretically, there is still a chance if the benefit/risk assessment in new settings is positive. In practice, this would have to be demonstrated before starting the pivotal trial at the level "There should be no doubt in the positive outcome even for the most critical reviewer." A big pharma company in such a situation may decide to terminate the development and cut the losses, since future development will require extra investment without hope of quick approval. (This actually happened to one of the HBOCs.) However, for a small, one-drug company it is the true "end of story". On the other hand, if you are able to demonstrate that the major culprit in your loss is *efficacy*, then your losses are limited to the unsuccessful pivotal trial. By the way, one important lesson learned: If one wants to finish a clinical development program successfully, then an RCT with a significant risk of a negative safety outcome should be avoided at any cost at the level of the design of a clinical program. Achieving this may be very challenging in practice. Some useful strategies include, but are not limited to: wise choice of clinical settings, populations, comparators, and treatment regimens, along with not rushing things and being flexible with final goals. Of course, the success is never 100% guaranteed.

Of course, there is no way to directly, accurately (and, most important, convincingly) separate results into "drug toxicity" and "drug lack of efficacy". However, there are a couple of approaches that could be used (preferably) together in order to make the case.

First of all, we have to make sure we have the case. The best indicator is the presence in our safety profile of the same AEs that are known to be in the safety profile of the treated condition. In our particular case, we were even able to separate our (well-known after 20 years of studying) side effect profile from the events that were typical for untreated anemia and were interpreted by reviewers as "significant safety concerns".

Note: Given the requirements for reporting of the side effects, all events observed after usage of the drug should be mentioned, and since there are no drugs with 100% efficacy, we have an interesting situation. If the treated condition has a MedDRA code, it is almost guaranteed that this condition will show up in the list of side effects! For instance, all recently approved drugs for diabetes have "diabetes" listed as a side effect.

The next step is logical, but (as the authors found) counterintuitive for a one-drug company: one may have to prove that the drug is *less* effective than the comparator. The authors did it by using a newly developed PK model of severe anemia treatment (that will be described in §6.2.3) and had all their belongings packed for a quick pick up before reporting it officially to their senior management. This might look like it was a risky gamble, but in fact it was a very pragmatic decision. Since the authors were confident in the presented results, in case of a rejection, the collection of unemployment benefits for a while was imminent (either immediately or after a short delay). This was a situation when the authors had to go against the current—after all, the words "we need an honest and truthful assessment" are often just words. Of course, it helps to immediately propose a way out (in our case, it was the change of the label) and make the case that this is the only possible solution.

The general assumptions for the assessment of efficacy in the treatment of severe anemia were based on the largely accepted ideas that the THb deficit and the time spent below transfusion threshold are descriptive of anemia severity, and the more severe anemia a patient experiences, the worse is the outcome. To the authors' great surprise, these assumptions were questioned despite a huge volume of publications confirming them in similar settings due to slight differences in other populations from the studied population. In order to get through this road block, we conducted analyses that demonstrated that this was the case in a specific study itself.

To make a long story short, it was an uphill battle that took 15 years, numerous statistical analyses of the collected data, several trials conducted solely to uncover a (nonexistent) mechanism of toxicity, tens of targeted publications to move away from the label of unsafe toxic drug. Frankly, this battle will be never won completely—there will always be a possibility to go back citing 10–15 year-old sources.

The main takeaway sounds like a common-sense trivia (see also the epigraph to §3.7):

It is much easier to avoid mistakes than to fix them later. And, yes, there are mistakes that are almost impossible to avoid.

Another takeaway from this experience is more disturbing: Sometimes rules and regulations may be in conflict with the logic of clinical development. If we are in a field that has not been adequately covered by existing regulations, we are on our own. Let us explain. The drug lost in a head-to-head comparison (in a close competition) to another drug that is a flagship in the field and is the only recognized player. An imposed clinical hold effectively prevents one for years from going into any clinical trial (possibly for another indication where no best drug is available). It is called "to err on the conservative side." The argument "If I am a high jumper and lost to the Olympic champion 2.20 vs. 2.40, would you permit me to compete against those with a combined best record of 0.50?" may not be convincing.

At this point, a careful reader may ask several valid questions, including the most important one: "Why did the company not engage early in a discussion with the regulators before undertaking such massive scientific efforts for two pivotal RCTs?" The modern drug development logic suggests that sponsors and health authorities are important scientific stakeholders (cf. Figure 1.2), and the earlier the dialogue starts, the better are overall chances for success. Many of the mistakes could have been avoided if the clinical development program was properly planned from the outset. Of course, this is similar to "looking at the answers in the back of the book." The nature of the scientific experimentation is never simple. Some of the arguments from the standpoint of our small pharma company are as follows: 1) Everything was ok before the start of the main pivotal phase III trial—no signs of potential problematic issues; 2) The main concerns from the FDA were rooted in toxicity demonstrated by other HBOCs in the field; hence there were subsequent FDA's recommendations to significantly lower the allowed dose; 3) The company and the FDA were not in close dialogue from the outset, etc.

Generally speaking, real life decisions can be of great complexity and often times cannot be made unequivocally. One example from the classic literature may be insightful here. Miguel de Cervantes' character Sancho Panza (who followed Don Quixote) once had to resolve an interesting puzzle as part of his duties of governor of an island. A law that governed the island prescribed that

when a stranger approaches the bridge to cross over to the island, he must
declare on oath to the guards on the bridge what will happen to him on the
island. If he swears truly, he is allowed to cross onto the island to pursue his
business; if he swears falsely, he is to be hanged. One day a stranger arrived
and coldly declared that he came to be hanged, and nothing else. The guards
were truly puzzled: on the one hand, if he is let to pass, then the stranger
lied—he declared that he came to be hanged; on the other hand, if he is
hanged, then the stranger told the truth, and by the law the judges should let
him pass free!

After summarizing the situation, governor Sancho's solution was: "of this
man they should let pass the part that has sworn truly, and hang the part that
has lied; and in this way the conditions of the passage will be fully complied
with." The person who sought the advice made a counter-argument: "But
then, senor governor, the man will have to be divided into two parts; and if
he is divided of course he will die; and so none of the requirements of the law
will be carried out, and it is absolutely necessary to comply with it." After
that, Sancho comes up with the moral solution:

"...there is the same reason for this passenger dying as for his living and
passing over the bridge; for if the truth saves him the falsehood equally
condemns him; and that being the case it is my opinion you should say to
the gentlemen who sent you to me that as the arguments for condemning
him and for absolving him are equally balanced, they should let him pass
freely, as it is always more praiseworthy to do good than to do evil."

This example may be instructive when we think about regulatory decision
making (e.g. Approve or Reject a new drug). Some new drug or biologic applications (such as in the HBOC field) may exhibit complex interplay between
risk and benefit and may require very careful considerations that are far beyond the binary Approve/Reject decision.

Now let us look at some hidden deficiencies and obvious mistakes that
eventually led to a failure of the entire group of clinical programs in this field.
To be absolutely frank to the HBOC developers, the authors had to admit
that some problems were inherited from the transfusion field and created an
extremely favorable situation for manufacturing of new ones.

6.2.3   Creation of a PK model for the transfusion field

*"Morpheus: This is your last chance. After this, there is no turning back.
You take the blue pill—the story ends, you wake up in your bed and believe
whatever you want to believe. You take the red pill—you stay in Wonderland
and I show you how deep the rabbit-hole is."*

The Matrix (1999) Movie

Looking back, it all started when a newly hired statistical programmer (with zero knowledge of the transfusion field, AP) while performing analyses on demand started asking foolish (by his own definition) and unnecessary (judging by reaction of those extremely busy with the BLA preparation) questions. When it became clear that those questions are mostly related to major hematology markers (THb, HCT, and plasma Hb), the director of pharmacology (BP) jumped into an opportunity to "check the basics", while educating the curious novice.

The first findings from this collaboration were interesting. It was well understood that HBOCs did not affect the RBC component of the blood, but rather dissolved in plasma. According to the regulations, the company had a well-developed PK model for an active ingredient of its product—free Hb measured in plasma. All results for plasma Hb in the pivotal trial were in agreement with the accurately calculated projections based on concentrations, half-life and volumes in circulation. But, this is where the good news ended.

The expectations of THb increase after one unit were based on clinicians' common knowledge cited in the protocol as 1 g/dL per unit and were a little off in the observed data (near 0.8–0.9 g/dL). For the HBOC infusion, the expectations were based essentially on the statement: "Since Hb content of 1 unit of the drug is twice as low as one unit of blood, the expected increase is 0.5 g/dL." This was confirmed by "napkin" calculations of increase as (Plasma Hb increase)×(1−HCT). These provided a close number, but both were not even close to the results observed in the pivotal trial (which were near 0.18 g/dL)—all per unit. The latest created panic among the principal investigators during the trial, which was left without explanation.

Note: The "napkin" calculations did not account for the fact that the infused unit of the solution had a significant volume (250 mL) and caused about 5% hemodilution; thus the starting THb value had to be adjusted by 5% down: for the starting Hb in the 6–8 g/dL range, the error was in the 0.3–0.4 g/dL range. The accurate calculations actually required much more effort and attention than the streamlined solution proposed later (because it worked separately with Plasma Hb and RBC Hb) but was obviously within the reach. The real reason why this was not done is the fact that the question was not formulated at that time. Later, when the first author got access to a BLA of another HBOC company, he found that the situation was widespread; moreover, the other company even did not use the "(1−HCT)" multiplier—they stated in their BLA that THb=RBC Hb+Plasma Hb, and used this "fact" for reporting of the PK results.

There was no model for the HCT besides general expectations that it should be lower in the HBOC arm throughout the treatment. (The observed difference was much bigger than the pharmacology colleagues would expect.)

What happened next deserves a thorough examination with some *amusing* conclusion, which is coming.

The novice, after getting all definitions of hematology markers had only one question: "Could the volume in circulation after transfusion be calculated as a sum of the volume before transfusion (V) and the volume of transfusion (v)?" After the affirmative answer, the basic formula for concentrations gave the increase in THb as $\delta = v\Delta/(v+V)$ (where Δ is the difference in concentration between transfused solution and starting THb), which immediately gave numbers that were very close to the ones observed for the drug and raised a question that led to some surprising revelations about the control (after we tried to answer the question of what happened here).

In short, nobody knew the exact volume and the concentration of the RBC solution, because nobody ever asked this question!

It took some time to explore the trial data to realize that the whole blood with more or less known concentrations (14–15 g/dL) and volume ($\approx$500 mL) was used in about 2–3% of transfusion, and the rest was the pRBC with much higher concentration, which, unlike volume (mainly between 250 and 300 mL), was not even captured in the data collection. It took much more efforts going through the literature and getting back to the sites to realize that: concentration of pRBC varies from 24 g/dL to 30+ g/dL, depending on technology; it is usually unknown to clinicians; and it is not normal practice to predict an impact of infusion on major hematology markers beyond the already mentioned rule of thumb: 1 g/dL per unit (which, by the way, can be really off mostly because of volume in circulation, since Δ for RBC does not change much for highly concentrated pRBC and starting THb near recommended transfusion trigger, say 5–9 g/dL).

Note: Here it is instructive to revisit our example in §1.5, where the task was to estimate the change in THb concentration after infusing 1 unit of RBC . This example is a good illustration of proper and improper use of statistics.

Of course, after using averaged values and this "magic" formula (and some of its variations), we were able to match all results from our studies for all hematology markers. In rare cases, when the reported results did not match, we were able to find programming errors, and after fixing those, we got a pretty good agreement with theory. (Note: it seems that correct predictions are powerful edit checks for produced reports!)

After all, there are no miracles or mysteries—applying simple math was enough to resolve the problem in this case. From a mathematician's point of view, the problem was solved without reservations, but surprisingly it received the opposite feedback—*unequivocal mathematical solution is not a proof for non-mathematicians!* (especially when they are used to live with a wrong solution). As an argument against the proposed solution we even heard a couple of times—"I am not buying this!"

Now let us formulate the promised amusing conclusion as a lesson learned:

Simple math that yields counterintuitive results must be sold.

The simpler the math is and the more counterintuitive the results are, the more difficult it may be to promote them. Ironically, the simplest way to promote them would be adding some more advanced tools (e.g. integrals, differential equations, etc.) even though they may be completely unnecessary. But no one would question sophisticated math!

In our case, we were forced to jump through many hoops in order to explain that the previously used dilution factor (recall: $v + V$ instead of V in the denominator) was overlooked and telling our audience that blood loss does not imply immediate drop in THb, whereas replacement of the lost blood by the volume expander does, etc., etc. We had to illustrate this with numerous pictures, tables and use some simple analogies to achieve acknowledgement. When a couple of years later we created an algorithm for long-term predictions that accounted for half-life of the drug in circulation and used differential equations and integrals behind the scene—there was not a single question about the created model! Our other lesson learned is: If you are fighting an established dynamic stereotype, use some sophisticated math if you want to have a fast resolution; or, if you use simple tools, be prepared for long discussions.

Here, the authors would like to add one more relevant quotation from their favorite character (Robert A. Heinlein "The Notebooks of Lazarus Long"):

> **The truth of a proposal has nothing to do with its credibility.**
> **And vice versa.**

What we discovered was quite disturbing. The company jumped into the pivotal trial without a relevant PK model, and without even a remote knowledge of the planned control. The only thing that the drug was actually expected to do—raising THb in anemia settings—the comparator did much better. The suspicion that the pivotal trial failed due to limited efficacy that started to float around during the BLA submission got stronger.

Later on, we found a much more disturbing fact: The entire HBOC field operated without such a model and it was inherited from the entire transfusion field. After a long meticulous search, we found only one(!) book that proposed to predict an impact of transfusion on major markers. None of the existing regulatory documents or SOPs established in hospitals ever mentioned such an "exotic" opportunity.

The next thing we had to deal with was the assessment of consequences of lower concentration and (as we found later, even more important) shorter half-life of our drug in comparison with the pRBC.

The existing dynamic stereotype was pretty simple: "So, what? Give more drug." When we tried to assess how much drug should be infused to achieve increases of THb comparable with 2 g/dL—which was a pretty standard expectation in severe anemia treatment—it became clear that we ran out of space in circulatory volume: the model started showing ridiculous numbers for planned infusions at the level of immediate infusions in the range of 3 L,

FIGURE 6.1
Safety analysis of HBOC trials

which would (at least) create serious cardiac problems for a normovolemic
patient due to fluid overload. Later we found that it actually happened at
the centers where clinicians tried to raise THb aggressively. The idea that we
just have to bridge patients for 6–7 days until their own RBC are reproduced,
applied to the model that accounted for circulatory half–life (less than one
day), gave astronomical numbers of the drug needed to complete treatment
while sustaining the same level of THb that pRBC would guarantee easily.
The arithmetics was simple: to fully replace 1 unit of RBC, we would have to
give 4 units of our product as the loading dose, and then give 2 units every
day just to sustain the achieved increase near 1 g/dL. The minimal assessment called for 10–12 units of the drug to replace 1 unit of the pRBC. Let
us add that the design of the pivotal study expected to use up to 10 units of
the drug to replace up to 6 units of the pRBC! As we found later, it was not
that bad just because the majority of control patients in standard practice
are over-treated; thus we had had some breathing room in our RCT.

The words "profound efficacy mismatch" in this situation look like an
overly optimistic evaluation.

Considering that our results gave ∼60% of blood avoidance with some
increase in mortality (3% vs. 2%) and in SAE incidence (22% vs. 18%), this
unsuccessful trial "dodged the bullet". Figures 6.1 and 6.2 can explain how we

FIGURE 6.2
Efficacy analysis of HBOC trials

looked at our safety outcome *before* and *after* the efficacy analyses. (Special thanks to the unknown author(s) of these great pictures!)

Now it is clear why we packed all our belongings before presenting these results at the company meeting. Our findings meant that:

- Due to restricted efficacy, our drug has a chance to safely replace pRBC only when requirements for pRBC are modest (up to 2–3 units).

- The entire treatment paradigm should be revised.

- The drug label probably should be changed.

- We have a long road ahead before any approval.

There was also some good news. We got a scientifically sound model that promised:

- To explain obtained results and predict future outcomes.

- To define weak links in our clinical development program and propose the realistic ways for refinement.

- To (finally) start working with clinicians on development of a new treatment paradigm that will be adjusted for the properties of our drug.

- In the long term, to aim our R&D to improve our drug using new formulations.

In short, we were not in the dark anymore. We got a very useful tool that permitted one to evaluate the situation in the entire HBOC field, and later even in the entire transfusion field.

We have to admit that there were two factors that helped us to survive an initial impact of these new findings and continue to work for the company:

- Higher concentration and longer half-life put our drug in the "top dog" position in the HBOC field (it was a pure coincidence—none of the authors was choosing the company using these criteria).

- Nobody, including the authors, knew how long and complicated our new path would be.

Final notes on the PK model: As George E. P. Box (1919–2013) once said: "*All models are wrong, but some are useful.*" Clearly, the presented model is oversimplification. There are many situations when this one-compartmental model is just plain wrong; e.g. after significant blood loss without any i.v. infusion, all available intravascular fluid would be pulled by the body in circulation. In fact, one of the main reasons for replacing volume is prevention of such situations. A description of these multi-compartment interactions is extremely complicated and, honestly, is impractical. Luckily, in 99% of the times this one-compartment model is sufficient. When adjustments are needed, they can be done, although they would still not be precise.

6.2.4 Mystery of the transfusion trigger

Preamble: When we started to work with our model, it became clear that if we want to maintain the level of THb that is typical for the pRBC treatment while avoiding fluid overload, the use of any other volume expanders should be severely restricted. Considering the fact that current practice calls first for volume restoration and then (after hitting the transfusion trigger) for Hb content restoration, we were forced to look at the history of the transfusion trigger, since we were preparing recommendations to increase it while using our drug.

Background on the transfusion trigger: There were times when the trigger did not exist numerically—blood was transfused, if restoration of volume did not resolve signs and symptoms of severe anemia. It worked well for the majority of patients, but created problems for *vulnerable* (compromised) patients (due to age, comorbidities, etc.), for whom the signs and symptoms should be avoided. In other words, the trigger was intended to keep "fragile" patients out of trouble before the troubles start instead of waiting for their manifestation.

There is no universal trigger for any population of patients, but there is probably the "right trigger" for each individual patient. The (absolutely) safe trigger for any subset of patients is the maximum of correct triggers for every patient in the subset. The price for safety is simple—all patients (the vast majority), for whom the correct trigger is less than the declared one, will likely be over-treated with blood.

The most practical approaches to stratify patients are based on two major factors: advanced age (say, > 65 years) and history of cardiovascular disease (the list may vary). The presence of any of these factors increases the trigger by 1 g/dL; the presence of both increases it by 2 g/dL. While the exact numbers are subjective, the use of similar gradations is common in clinical practice around the world.

There is a perception of some unequivocal *safety threshold* of THb among clinicians, which guarantees uneventful treatment of severe anemia that usually serves as the *target THb*—patients should reach this level fast and then it should be maintained. According to available data, the safety threshold is more or less stable despite improved clinical practice, whereas the transfusion trigger in the recent 20–30 years has a clear tendency to go down. The transition from whole blood to pRBC started in the 1970–80's and it was practically completed before the 1990's.

Mathematician's notes: The attempts to rigorously calculate the transfusion trigger (or the safety threshold) for a specific patient are obviously futile. Any black box modeling is doomed not only because of the big number of covariates, but simply due to the fact that even if the correct number exists, it cannot be found. All attempts to refine the trigger in RCTs turned out to be inconclusive. In general, outcomes were similar for the "restrictive" (more respect to the symptoms) and "liberal" (more respect to the numbers) treatment strategies, with two common results for all such RCTs: Significantly less blood was used in the "restrictive" arm, and in some subgroups there was inferior safety in comparison to the "liberal" arm. We will revisit these results later.

Additional details: There were a couple of items that got our attention:

- Fluid overload that we have already seen in our data was listed in one very old transfusion guidance as a common side-effect of the whole blood while treating severe anemia in normovolemic patients.

- The transfusion trigger and the safety threshold were almost identical in the 1990's, but then the trigger went down (around the world) and stabilized, on average, at 1.5–2 g/dL lower than the safety threshold, which remained unchanged. Recall that we defined the *safety threshold* as a level of THb at which anemic patients are usually kept after initial/satisfactory treatment.

Thinking process: Frankly, we were looking specifically for the first item (fluid overload), because it was our prediction after applying the PK model. The fact that it was the prediction of the past events did not diminish its value—it actually confirmed our modeling.

The second item became clear after some thinking and consultations with clinicians. It seemed that the acceptable/comfortable transfusion trigger for clinicians had to be within striking distance of the safety threshold. On average, 0.5 L could be transfused immediately without risk of overload for almost any patient. The expected increase for whole blood is about 0.4 g/dL, which is measurable but hardly accounted in perception (which serves as the psychological base for the trigger). Raising THb by 2 g/dL is a long and painful process based on the fact that the plasma is eliminated much faster from circulation than the RBC. We had a strong suspicion that it does eliminate faster (if there is extra volume in circulation), but it was very hard to prove. Despite this, we had at hand a very plausible (and confirmed) hypothesis.

The other thing that old manuals confirmed was our suspicion that times when the whole blood clinicians had to be masterful in simultaneous restoration of volume and THb content became obsolete once the pRBC became available. No longer would whole blood be needed as long as a high-concentration solution is available. Common knowledge among physicians is that big pharma forced the use of separated blood components into clinical practice; however it is clear that it was readily accepted because it simplified treatment enormously (cf. our example about the Sun and the Earth from §1.5). Ironically, with treatment of major blood loss due to trauma, history made a full circle. All recent attempts to find a proper combination of components in these settings are finally converging to combinations that mimic whole blood.

Now that we have described *how* we reached the solution, let us formulate our answer to the *mystery of the transfusion trigger*.

Transfusion trigger as a reflection of efficacy of used oxygen-carrier: We have already discussed that the transfusion trigger, which was established during years of whole blood usage, was much higher than the modern recommendations are. Historically, after switching to the pRBC, it stayed at the same level for a decade or two, and then started to go down. The proponents of these changes (and future decrease) conducted observational studies and even RCTs to demonstrate that the safety outcome does not depend on the treatment strategy—usage of 10 g/dL does not provide extra benefits compared to the 8 g/dL threshold. The results are used to support the point that many physicians, especially cardiologists, cannot agree with: the safety threshold for anemia is below 8 g/dL, and even closer to 5–6 g/dL. Furthermore, some studies even demonstrate that healthy young individuals can tolerate 4–5 g/dL without any serious consequence, which, according to the proponents, supports the modern trend for lowering the transfusion trigger.

The opponents of this trend would argue that there is a significant sub-population of "fragile" patients (elderly; those with history of cardiac problems; etc.) who require a much higher transfusion trigger to assure their safety. Some studies provide evidence that the safety threshold for "fragile" patients could be as high as 11–12 g/dL. Moreover, a part of this sub-population cannot be detected by formal profiling, thus lower transfusion triggers introduce unnecessary risk to them. Current recommendations represent a compromise between two positions: they require transfusion below 7 g/dL; do not recommend it above 10 g/dL; and leave a grey area (7–10 g/dL) to the physician's discretion.

Some guidelines for the transfusion field go further, by proposing 8 g/dL with adjustments of 1 g/dL for both age and history of cardiac disease (which makes 10 g/dL a proper trigger for an old patient with a history of cardiac disease). Nobody believes that there is such a thing as a universal transfusion trigger, but it appears that in this ongoing discussion, all parties almost agreed that the proper transfusion trigger and the safety threshold are generally the same. Strictly speaking, the safety threshold is a level of Hb where you want to be, whereas the transfusion trigger is a level of Hb at which you start acting in order to reach the safety threshold. The difference is one's realistic evaluation of their ability to increase THb immediately. Considering one unit of whole blood as an immediately available response, in the proximity of 10–11 g/dL one has an increase in the range 0.3–0.4 g/dL. Of course, in such a situation the transfusion trigger and the safety threshold had to be essentially the same. When pRBC became available, the situation changed dramatically—under the assumption of the same volume for immediate response, we have at our dispense 2 units of pRBC, with an impact on THb in the range 1.5–2 g/dL. No wonder that the safety threshold has remained unchanged whereas the transfusion trigger has gone down.

These numbers describe a "normal" situation—reasonably stable patient conditions and a flawless transfusion decision. In critical situations, due to fast changes or errors in the transfusion decision, it gets even more complicated.

Let us assess a possible impact of a "rescue operation" on THb near 8 g/dL (which is the modern transfusion trigger) in different settings (whole blood vs. pRBC). A simple model that accounts for volume restrictions and our ability to overcome them (e.g. diuretics usage, clearance of previously infused crystalloids, etc.), sets limits for THb increase in the 1–1.5 g/dL range for whole blood, and in the 4–5 g/dL range for pRBC. These numbers define "points of no return" relative to the true safety threshold.

Obviously, the pRBC environment is much more merciful to errors, miscalculations and just emergencies. For instance, a miscalculation by 2 g/dL safety threshold in the whole blood environment would leave a patient under safety threshold for days, if not weeks. On the other hand, the same error in the pRBC-controlled environment could be corrected in minutes. The latter actually explains why modern transfusion triggers being overly optimistic do not influence safety significantly—they are still within reach for immediate correction when/if needed. Of course, there is a price to pay: there is no room for true emergencies or big miscalculations, but the difference is too small to be detected in observational studies. Even with up to 20% rate of emergency/errors and up to 20% rate of delayed corrections, the difference in observed safety outcomes would be up to 4% (20% × 20%). In order to statistically detect such a difference in a well-controlled clinical trial, thousands of patients would be required.

We do not believe that this trend will continue, especially in a situation when physicians already do not pay too much respect to existing triggers. The reports of level of THb before first transfusion released by different hospitals show that for significant (> 20%) fraction of patients, THb was above 9 g/dL, and the level of THb that all patients were maintained at was 1–2 g/dL higher than THb before the first transfusion.

Summary: Current transfusion recommendations are already overly optimistic and potentially endanger a small (1–5%) fraction of patients. This is widely offset in clinical practice by the use of non-THb-based indications for transfusions. If transfusion triggers decrease further, the risks will increase significantly. Safe usage of whole blood under the current treatment paradigm is limited to direct replacement of blood loss with consequential pRBC correction when the blood volume is restored. Hypothetical attempts to get back to the good old whole blood under the existing standard of care will have disastrous safety results. Whole blood would lose to pRBC in clinical trials, even with adjustments for the standard of care—this is exactly what happened when pRBC replaced whole blood 40 years ago. A clinical trial using the existing standard of care would end up with declaration of whole blood unsafe and probably toxic. There would be numerous publications "explaining" mechanism of toxicity of whole blood and whole blood would eventually be completely banned for usage in order to protect patient safety. The statement that whole blood is an ideal solution for blood loss replacement would be deemed as old-fashioned, non-scientific, and naïve by some scientists.

6.2.5 The rise and fall of the HBOC field

"Happy families are all alike; every unhappy family is unhappy in its own way."

Leo Tolstoy "Anna Karenina"

If we transfer the epigraph above into the field of SOPs and regulations, we may say that all covered situations are typical (just because they are already well-classified and generalized), while non-covered situations are unique. When they are encountered enough times (to recognize typicality), they can be articulated in SOPs and regulations.

There is probably not (and hopefully never be) a field in drug development that is better suited for analysis and summary of mistakes that led to failure of clinical programs. Despite profound differences in formulations, structures of the programs, development times, sizes of companies, actual efficacy of drugs, etc., all of these programs and companies followed similar paths of destructions. In hindsight, it seems that all problems, whether they were objectively present from the very beginning or created by companies, could have been avoided. Two of the authors (AP and LBP) that have combined 25 years of work in this field feel obliged to, at the minimum, share their observations, conclusions, and generalizations.

The common scenario was:

- A successful pre-clinical program confirming everything that had to be confirmed: exceptional efficacy and satisfactory safety with modest side-effect profile.

- Convincing phase I studies demonstrating that the drug works as intended and actually treats the condition described in the label.

- Satisfactory phase II studies confirming side-effects and observing some safety signals that were ruled as non-significant.

- Failed phase III study (studies) with the results varying from near miss to disaster.

So, what were the reasons and how could they be avoided? The first set of reasons is grounded in non-standard properties of these biologics that create objective difficulties in development, since they are not covered properly by existing regulations:

- Dosing based on need.

- Different levels of efficacy:

  - "top dogs" ($[Hb] > 8$ g/dL) were in the worst situation since they were real oxygen-carriers and potent volume expanders at the same time;

    – the situation was a little better for "underdogs" with [Hb] < 5 g/dL
      (that have negligent or even negative impact on THb near the transfu-
      sion trigger) since they were forced to look for a new indication instead
      of anemia treatment at early stages.

- Absence of a relevant PK model for the anemia treatment: we put it in
  the "objective" set because it was inherited from the transfusion field.

We have already discussed in detail the first and the last bullet points, and
in order to understand complications of the second one, we need:

Severe Anemia 102

Severe anemia is a very complex condition and is treated by many different means, including different types of drugs that serve different purposes.
Probably, the three major goals that might be achieved with i.v. infusions
are: 1) restoration of circulatory volume; 2) restoration of the THb content
and concentration; and 3) maintaining of coagulation properties of blood in
circulation.

The efficacy of solutions that are currently used as the standard of care
cannot be compared as *efficacy in general*. The only product that addresses
all three goals is good old whole blood, but it also has deficiencies (due to
relatively low Hb concentration) and, more importantly, it is rarely used in
hospital settings. *Crystalloids* and *colloids* are used solely for restoration of
circulatory volume and even they cannot be compared directly since the former
provide short term (minutes) restoration, whereas the latter have more prolonged effect (hours or even days). The pRBC are poor volume expanders and
are almost deprived of coagulation properties, yet they are extremely efficient
in raising THb. Plasma products are good as long-term volume expanders and
maintain coagulation. The HBOCs are potent volume expanders that act as
colloids and have, on top of this, varying oxygen-carrying abilities. The "un-
derdogs" ([Hb] < 5 g/dL) tended to find their niche as volume expanders and,
in reality, they were never tested in demanding settings of life-threatening
anemia (essentially, they struggled to show how they were better than the
much cheaper colloids). The "top dogs" ([Hb] > 8 g/dL) were tested either
against the pRBC or (as researchers thought) against crystalloid/colloids.

It may seem like nothing special—pick up your niche, conduct an RCT
against the comparator and get all your answers (which, by the way, was
done). The problem is that in the settings of interwoven multiple efficacies,
designing RCTs that would give answers to the questions you want to be
answered is extremely hard. As we have already seen, going against the pRBC
was a recipe for disaster. This raises a couple of questions.

Question 1: Why was efficacy deficiency (of HBOC-201 that was taking
a high road of replacing blood) not uncovered pre-clinically?

The answer is stunningly simple: All pre-clinical testing was done against
(animal) blood, where HBOC-201 was superior in oxygen-carrying abilities—

13 g/dL vs. 8–10 g/dL. It is quite possible that the results of the RCT would have been much more satisfactory than they actually were if HBOC-201 was tested against (human) whole blood with concentration near 14 g/dL; but nobody realized during transition from pre-clinical to clinical that now the comparator is the pRBC with concentration of 24–30 g/dL.

The lesson is clear: *The comparator(s) for a clinical program should be not only chosen wisely, but in addition checked thoroughly.*

The other lesson is: *Transition from pre-clinical to clinical is commonly a weak link, where continuity of the entire program can be easily broken.* (Possible reasons: different approaches, institutions, researchers, regulatory requirements, etc.)

Question 2: Why did the "top dogs" in the HBOC field not perform overwhelmingly well against the crystalloids/colloids? (In fact, they lost all pivotal trials, with varying convincing degrees).

Let us put aside for now the putative toxicity of these drugs, because the SAE profile was (at least partially) similar to the untreated anemia, which means that we are potentially looking again into the efficacy issue.

Again, the answer is simple: *There was no single trial in severe anemia settings where the HBOC competed solely against crystalloid/colloids; the element of competition with the pRBC was inherently present, whether researchers planned for it or not.*

Let us look at the pictures that illustrate what happened when one steps up into an already existing standard of care tailored for a highly concentrated solution, when the new drug is low-concentrated and one does not adjust the treatment paradigm.

Figure 6.3 explains what happened in a direct competition with the pRBC in two pivotal trials (114 and 115) with HBOC-201. The design permitted a limited (absolutely insufficient in hard cases) number of units of HBOC to be infused; thereafter, if it was not enough, the patients were crossed over to the gold standard of care—the pRBC. One study permitted 7 units of drug, while the other one permitted 10 units. The rationale behind these numbers was as follows. It was more or less proven (using the methodology that we are omitting from the discussion here) that 7 units of HBOC, on average, permit blood avoidance with the actual need of 2 units or less of pBRC (114 trial), while 10 units permit safe substitution of 3 or fewer units (115 trial).

We looked at what happened to patients who needed 2 (3 for the 115 trial) or more units of the pRBC, to begin with, and even theoretically could not be treated solely with the allowed HBOC infusions. The patients in the control arm reached the safety threshold within the first several hours of treatment, while those in the HBOC arm reached this threshold only after being crossed-over to the pRBC (with a delay that averaged 3 days in the 114 trial and 4 days in the 115 trial). Roughly speaking, all imbalances in safety are attributed to this delay. Note that it would be unfair to say that the HBOC did nothing for

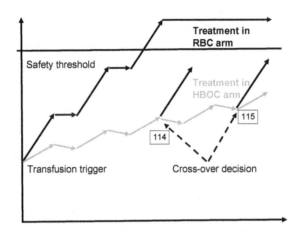

FIGURE 6.3
Hb content over time in RBC-controlled trials in cross-over groups

these patients—if they were treated only with volume expanders, their THb would have gone much lower, and the results would have been much worse. As we already noted, we dodged the bullet in this trial because with the standard of care, patients are commonly over-treated with pRBC transfusions.

Figure 6.4 demonstrates a typical situation in the so-called *crystalloid/colloid* trials. The HBOC patients during the (usually very short) CTM part would have an "upper hand", but then for a much longer time they would be in the "grey zone" between the safety threshold and the transfusion trigger, while the control patients would be above the safety threshold almost immediately after the CTM part is completed.

Unlike the first situation, which requires dramatic changes in the treatment paradigm to achieve safety in comparison to the control, the second situation permits a quick fix that would assure an "upper hand" for the HBOC arm during the prolonged observational part: physicians should just use with the existing trigger the RBC Hb instead of THb, thereby ignoring a short living part of THb.

Given the fact that the discussed scenario describes a standard trauma trial (where HBOCs are planned to be used), this observation has not only theoretical but also practical value. The last pivotal trauma trial with the HBOC Polyheme has shown that this explanation is, at the very least, plausible—mortality in the Polyheme arm was lower (about 3.5% in absolute numbers) until THb was higher (first 2–3 h); then during the period when THb was higher in the control arm, the situation was reversed, and in 12 hours the difference in mortality stabilized at roughly the same level, but in favor of the control. Probably, the results would have been different if this factor was accounted for during the design of the study.

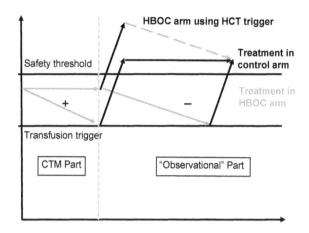

FIGURE 6.4
Hb content/concentration in the crystalloid/colloid trials

Takeaway:

1. We should not forget that an RCT tests not just a dose of the new drug, but also the way the drug is used. It is our responsibility to develop and test an appropriate treatment paradigm, which might vary in different settings.

2. We should not forget that the new drug is often introduced on top of the existing standard of care. If the standard of care needs modification(s) to accommodate the use of the new drug—we should change it, or adjust the treatment paradigm, or both.

Next are two more examples from the HBOC field that illustrate these recommendations:

1. If massive infusions of the HBOC are expected, *dilutional coagulopathy* may become an issue. This was an additional, unaccounted for problem in the Polyheme trauma trial.

2. The standard of care uses blood pressure as the main marker of volume status. All HBOCs are vasoactive. Reliability of blood pressure turned out to be an issue only during transition to pRBC, which is not vasoactive. There is a growing suspicion that during the CTM part *resuscitation* is achieved with lower volume; thus during this transition there is a risk of hypovolemia. Presumably, it was one of the reasons for disastrous results of the trauma trial with Hemassist, which was the most vasoactive HBOC ever tested in an RCT.

Let us highlight some other mistakes that are unfortunately very common:

- Ignoring or downplaying negative findings in early stages of development.

- Upgrading small positive trends to the status of established achievements.

- Blaming "bad luck" during randomization for the poor outcome of the RCT.

- Non-admission of one's own mistakes, which translates into non-learning.

6.3   Summary

It is impossible that all drug development programs succeed—there will be both successes and failures. What is important is to maximize the chances of correct decisions at different stages of development. Avoidance of all mistakes in non-standard, non-covered situations is wishful thinking. What is realistic is the detection of problems ASAP (never in time in practice!) and attempts (by trial-and-error, starting from resolution and negotiation with the health authorities) to address them.

There will always be an element of unpredictability in any open-ended project (which we discuss in Chapter 7). Also, we should ensure that statistics as a tool for new knowledge acquisition (evidence-based science) is used properly, while ensuring compliance with the regulatory standards to protect patients (present and future). After all, most of us will be patients one day.

Let us focus again on the regulatory requirements. What is the purpose of regulations? In theory, it is to develop some rules (guidelines, SOPs, etc.) to ensure harmonized, scientifically accurate, reproducible results. It works well as long as the rules adequately cover all situations. However, in practice, this is not realistic. We already mentioned one good example when an expert working group (regulatory and industry statisticians) was formed to amend the ICH E9 guideline [61] by providing a structured framework of the estimands and the sensitivity analysis. The resulting document ICH E9(R1) [58] was developed in 2017 and released for public comment. Similar logic is used by health authorities to provide guidelines on other complex (statistical) issues, such as adaptive designs, multiplicity, master protocols, to name a few. This highlights the importance of global collaboration in such a complex enterprise as the drug development. The triangle health authorities–academia–industry provides a very powerful platform for the open scientific discussion.

The authors do acknowledge that all their conclusions and recommendations presented here are subjective and highly opinionated. The intent of this overview of potential problems and proposed (when possible) remedies is to bring attention to some important topics, illustrate application of mathematical and statistical skills in complex problem solving, and highlight the importance of transparent scientific discussion among various stakeholders in drug development.

7

Can One Predict Unpredictable?

> *"...You can't always get what you want*
> *But if you try sometimes you just might find*
> *You just might find*
> *You get what you need."*

The Rolling Stones. Songwriters: Keith Richards/Mick Jagger.

7.1 Personal disclaimer/preamble

The first author has believed all his life that he clearly understands the term "realistic and honest planning" in all situations, starting from many-times repeated tasks under absolutely stable conditions, and including the projects we have never seen before and have no idea whether they can be successfully finished at all. In some sense, in real life the minimal skills in realistic planning are paramount for survival—while we normally would not give any thought to the task of walking from bedroom to kitchen in our own house, we have to think seriously about getting across a busy highway. Undoubtedly, even though the meaning of the word "planning" changes considerably from task to task, it still remains planning: we are planning to achieve desirable results, irrespectively of whether we do it consciously or not. In this chapter we shall focus on *business* planning, while keeping personal planning at hand to highlight some standard problems that exist in business planning.

In short, the first author has been convinced that Murphy's Law is just a logical reaction to overly aggressive business planning. The formal answer to the question "What distinguishes proper planning from the one commonly used?" definitely required a lot of work and time; thus it was postponed indefinitely with minimal chances to ever get back. The situation changed in 2009, when the first author was happily unemployed (when the first incarnation of the Hemopure manufacturer went bankrupt) and got a job proposal from a software company to create a scientific base for their planning tool that would help with risk management and contingency plans by detecting projects that

159

were going nowhere in terms of profitability. It was almost immediately clear that more or less accurate planning of timelines for open-ended projects was a pre-requisite for tackling a bigger problem. Moreover, from the mathematical point of view, these two problems are equivalent: if planning is done accurately for any complex task, the rest is relatively simple. After clarifying the non-interesting (for this chapter) part of the project during the first month or two with the employer, the author finally achieved the ultimate goal of any researcher: the possibility to satisfy one's own curiosity while being paid for these efforts. The following chapter discards all economical parts and focuses on pure planning of the timelines. Let us try to understand the problem a little better.

7.2   First, what can we do?

There is no problem whatsoever in planning of the well-known, repeated many times processes, no matter how complicated they are. After all, mass production, which is the core of civilization, is based on this ability. We can call it *deterministic* planning. The main features are well-studied, and the entire approach is excelled to perfection. In essence, it is a creation of a *technology map*: the entire process is broken into elementary tasks which are organized along the required sequence of execution. Some of these tasks can be performed in parallel; others should be done sequentially. If time for execution of each elementary task is known, we can find the so-called *critical path*—the shortest possible way to completion of the project, assuming that any parts that can be done in parallel will be done in parallel. It is undoubtedly one of the most important parameters in planning, but it has a couple of features that are not always clearly articulated:

1. The theoretical critical path is almost never optimal. For instance, we have two parallel tasks at some point that will require extra machinery and/or a qualified work force. It will never pay for itself in a single project, and not always will it be useful even in mass production.

2. Both the critical path and the optimal path heavily depend on the developed technology map of the entire project/process, especially on the content of elementary tasks.

3. The technology map can change due to an enormous number of factors, including changing priorities, availability of resources, number of repetitions, etc.

Let us add some observations about elementary tasks. They are usually analyzed by breaking down the process, preferably to the level of single simple operations, and then re-synthesized into optimal building blocks, on which the technology map is built.

In addition, we may note that any technology map can be structured using multiple levels for building blocks; thus any sub-project can be viewed as an elementary task for a (sub)project of a higher level. In well-organized technology maps, planning for elementary tasks is extremely tight—if there are many different tasks and the number of repetitions is high, then for a flawless execution you really need the Six Sigma technique.

In the best traditions of problem solving, we shall assume from now on that this part of planning is already done and can be repeated again any time when we need it, which means we can move to real problems that exist in planning and focus on them.

7.3 Problems in planning of the open-ended projects

"Pessimist by policy, optimist by temperament—it is possible to be both. How? By never taking an unnecessary chance and by minimizing risks you can't avoid."

Robert A. Heinlein "The Notebooks of Lazarus Long"

First, let us give a formal definition of an *open-ended* project. It is the unique project that has never been done before and contains at least one element with some degree of uncertainty: ranging from an imprecise timeline needed to execute the project to an unclear outcome.

The word *unique* is important: If you repeat a very complex, but *exactly the same* project that was successfully executed before, there is still an incredible amount of uncertainty and you might be quite off with your planning, but strictly speaking it is not open-ended. The more runs you have, the more precise your planning will be. Moreover, the technology map will improve dramatically after 2–3 runs. At the same time, the words *exactly the same* are equally important. For example, any clinical trial (let alone clinical development program) is an open-ended project, even for a huge CRO that completed thousands of them, just because it is unique and includes elements of uncertainty that cannot be removed completely. Here we are not talking about predicting the results (which is impossible, at least, with 100% confidence), but "just" about completion of the trial. Of course, previous experience with similar or not so similar situations will help tremendously, but there are many things outside of the planner's control that can prolong, or even completely derail the project. A very limited set of possible hurdles includes: slow enrollment, unpredicted negative outcomes, poor execution, lack of financing, changing priorities, etc. In the IT industry, similar open-ended projects are projects to develop some new software packages.

Both IT and pharma industries are notorious for the rate of failures and underestimations of the needed time and money, just because their critical building blocks are mostly open-ended projects. Of course, these two industries are runners-up—the gold medal belongs to Big Construction. After all, the Cheops Law: "*Everything takes longer and costs more*" is probably older than pyramids.

A major killer of open-ended projects is the presence of the planned part that has not been completed. Not necessarily does it mean "the end of story"—there are many situations when the planned part that looked like an absolute "must do" could be replaced by a sequence of parts that can bypass a road block or even streamline the process tremendously. In the majority of cases, though, "it takes longer and costs more", which usually happens when the planned part from the beginning was just a brave shortcut that bypassed re-alistic planning. (Note an interesting difference between bypassing an *obstacle* and bypassing *planning*.) Surprisingly, there are many situations when some tasks, besides being impossible to complete, are absolutely *unnecessary* for a successful completion of the project. The other side of same problem is the presence of *necessary* parts that were overlooked during the initial planning—the consequences in this case can be much worse than in the case of unnec-essary tasks. This issue is so important and disregarded so frequently that it deserves special discussion before we can move forward.

7.3.1   Extraneous vs. overlooked parts in preliminary planning

> "...*The brain is cleverer than that:*
> *It was my first adjustment that was wrong,*
> *Adjusted to be nothing else but brain;*
> *Slave-engineered to work but not construct.*"

John Wain "Poem Feigned to Have Been Written by an Electronic Brain"

Since *overlooked* parts usually have disastrous consequences, they are eas-ier to explain/understand (which in no way means that they are easy to avoid). Some common examples in drug development include:

- A missing study (either non-clinical or clinical) in a development program that leaves some critical questions unanswered and has a potential to derail the entire program.

- A missing part in a planned study, e.g. data collection that is necessary for the planned interpretation of the results, which cannot be restored by revisiting all collected data.

It is easy to say that one should understand all parts of the planned project in advance, but it is incredibly hard to implement this consistently throughout initial planning—chances are that something will be missed. The only practical recommendation in this situation is to assure that inside the omissions there are no elements that are absolutely critical for project completion.

Now, let us look closer at the *extraneous* parts. Their presence is linked to many reasons, most frequently an unclear understanding of the problem at hand, coupled with going by existing stereotypes that worked in similar situations, which in some case means (apologies for blasphemy) following existing SOPs. In reality, these extraneous parts are discovered only under extreme pressure, such as the mentioned inability to complete them.

In order to demonstrate what we are talking about, let us analyze a simple math problem. Someone is rowing on a boat (with a speed of V m/sec) along the river against the current (whose speed is v m/sec). At time T, this someone drops his hat into the water, but notices this only after t minutes have passed. Immediately after noticing this, the person turns around, doubles his own speed and starts rowing along the river with the current. The question is: How long will it take to retrieve the hat?

A standard solution by a meticulous researcher would likely include at first the assessment during time period from T to $(T + t)$: the rower is going with the speed $(V - v)$ in one direction, while the hat is going with the current speed v in the opposite direction; then at time $(T + t)$, how far the rower will travel: $(V - v)t$, how far the hat will travel: vt, and what is the distance (S) between them at the moment when the rower turned back: $S = (V - v)t + vt$. Then the speed of the rower will be $2V + v$ and he will gain on the hat that continues down the current with the speed v, with the speed $(2V + v) - v$. After dividing S by this difference, the desired answer is obtained.

If the researcher works with symbols, he may find out that the answer does not depend on V and v and equals exactly $t/2$. If not, then the only way/reason/motivation to find it out is unavailability of this information. The distance $S = Vt$, together with the fact that at first the rower departs from the hat with a relative speed V, and later returns with the relative speed $2V$, might provide a thoughtful researcher with clues about what is really going on, and why all these perfectly logical steps (that represent a couple of unnecessary elementary tasks) are extraneous for the entire problem. The true nature of the task at hand will become clear if we replace the river with the moving train (or, following the great people like Galileo or Isaac, we can replace it with the Earth): the irrelevance of the speed of the train (or the Earth), as well as the speed of someone going for the dropped hat, becomes obvious.

This particular situation is actually circumstantial and psychological, which provides some interesting practical recommendations. Suppose we are requested to perform some tasks A, B and C, and after a thoughtful question "Why would we need this done?", we learn that this serves a higher purpose to achieve D. Now, suppose we see that steps A and B could be dropped after

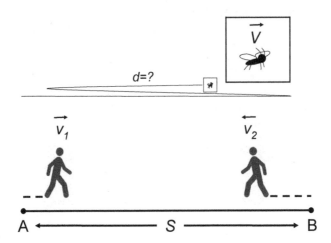

FIGURE 7.1
What distance will the fly travel?

a slight modification of C, which permits doing D much faster. If A and B are standard tasks and our modification of C has novelty, in 99% cases we may be out of luck trying to explain that a better approach exists and the strategy should be changed. We actually have better chances if we can demonstrate that tasks A and B are either impossible to complete or prohibitive due to time or resource constraints. Unfortunately, if we must operate in accordance with the existing SOP, there is practically no choice—just do A, B and C to get D, although it may take very long and/or incur high cost.

We know only one way out in practice: If A and B are truly "impossible" to complete, pretend that you do them both and later use the results (which are actually the product of imagination and creativity), while in fact you silently execute a modified C to get D. Just do not tell anybody! This entire situation is usually pretty harmful even in a normal scenario, which means that due to lack of understanding, we often do a lot of unnecessary steps that sometimes require efforts of a much higher magnitude and a much more superior toolkit than a streamlined well-thought out solution.

Let us take a look at another powerful example. Two pedestrians are moving toward each other from points A and B, respectively. The distance between A and B is S and the speeds of the pedestrians are v_1 and v_2, respectively. There is a fly that is moving much faster with speed V, which starts with one pedestrian, goes to the other, then goes back to the first one and so on, until the two meet each other (cf. Figure 7.1). The question is: What distance will this fly travel?

The added value of this example is a demonstration of how (dynamic) stereotypes are created. There is no problem whatsoever to find out what happens at the first leg of the flight: the time t_1 is $S/(V + v_2)$; the fly traveled Vt_1, and the distance between the pedestrians is now $S_1 = S - v_1 t_1 - v_2 t_1$. The calculations for the second leg are not more complicated: the time t_2 will be $S_1/(V + v_1)$; the fly traveled $V(t_1 + t_2)$, and so the distance between the pedestrians is $S_2 = S_1 - v_1 t_2 - v_2 t_2$. It is obvious how to proceed further, and after 3–4 iterations, we may consider writing an SOP, and after 10–15 repetitions it could become a "best practice". The main problem with this approach is that the number of legs is infinite. A practical solution could be to restrict the number of legs—sensibly limit it based on the remaining distance between the pedestrians (say, 1 meter). The theoretical solution is much more complicated and involves infinite summation—essentially one will need to recognize a geometric progression; otherwise it will take an even more sophisticated technique which involves differential equations and future integration. Of course, the question of how many times the fly will change directions will remain unanswered, but it might as well be required to report in study documentation.

By contrast, let us look at the streamlined solution: The pedestrians will need the time $T = S/(v_1 + v_2)$ to meet; all this time the fly will travel with speed V, thereby covering the distance $VT = VS/(v_1 + v_2)$.

Note: Make no mistake—no matter how logical this solution is, if there are SOPs in place to make 100 iterations, we may have a hard time selling the solution! If we try to publish it first, one reviewer may call us "naïve" while another may accuse us of making a forbidden shortcut or oversimplification. (This is a joke, of course).

Let us go back to the planning of open-ended projects. Assume that we already have all elementary tasks evaluated. We need to eliminate unnecessary parts of two kinds: those that are non-achievable, and those that are "nice to have". Ironically, getting rid of the first kind is a must, while getting rid of the second kind is just (again) nice to have.

It is actually common knowledge that we are not planning to obtain major discoveries inside our open-ended projects; nor are we planning for impossible or unachievable goals. The last part is related to the difference between "marketing" and "realistic" planning. There are two types of consequences of an overly brave "marketing" planning:

- "Just" underestimation of time and cost needed.

- Absolute inability to deliver the final goal.

The difference is in the nature of the planned task. Let us use some simple examples to illustrate the first bullet point. Suppose our best result in running one mile in high school (when we were 20 years younger and 50 pounds lighter) was 5 minutes. We are fooling ourselves not only when planning for less than 5 minutes, but probably if planning for less than 6 (or 8) as well. It is just an "optimistic" underestimation, but it may become a deadly mistake if we are preparing to jump over an abyss keeping in mind our high school record of 6 yards. Jumping over anything more than 5 yards will be suicidal, and considering the high stakes, we should probably never plan for anything more than 3. Unfortunately, in reality it is not uncommon to observe "very aggressive" planning that uses as "reasonable" goals the best result achieved by the best athlete in our high school over the entire period of 20 years since we graduated.

Even after eliminating these issues, we need to discuss the *level of uncertainty* of elementary tasks.

7.3.2   Level of uncertainty of elementary tasks

I. No uncertainty at all

This is a scenario when we know the task pretty well and know the exact time (given availability of resources) needed to complete the task. The task without uncertainty could be dealt with using deterministic planning—it does not represent any special interest to us. Note that absence of uncertainty is not absolute and depends on the planning unit of time; e.g. if we know that a task requires between 2 days+2 hours and 2 days+6 hours to complete, for the planning that uses "day" as a minimal unit, the task will require 3 days. But if our planning unit is "hour", we will have an uncertainty (say, between 18 and 22 hours). Moreover, it may depend on specific properties of the estimate as well—if we expand an elementary task by adding a subtask that takes exactly 4 hours, we may suddenly get uncertainty even if we are planning in days, because our estimate will be between 2 days+6 hours and 3 days+2 hours, which will translate into 3–4 days.

II. Minimal uncertainty

In this scenario we have a pretty good idea about realistic boundaries of time to completion, and these boundaries are tight enough to not disturb the critical or optimal path that we follow. These uncertainties are most promising to deal with using standard statistical approaches. The key is the knowledge of an (approximate) distribution of time to completion for such tasks. Even (or especially) in a long sequence of such tasks, we have a chance for a reasonable estimation of time to completion; indeed, if these tasks are independent, the mathematical expectation for mean time to completion is a simple sum of mathematical expectations of mean times for each task and the Central Limit Theorem (CLT) will take care of the 95% confidence intervals by not permitting them go too wide. After all, if we have random deviations from means, they should compensate each other. In some sense, this idea is the Holy Grail of planning that accounts for random deviations. The problem is that it obviously will work only for a very limited class of distributions; e.g. if we have even a couple of bimodal distributions in our sequence, going by mathematical expectations loses any sense. It is easy to show that, depending on the numbers, we may end up with 3–4-modal distribution (where 3 is a lucky exception, if the left mode from the first distribution + the right mode from the second one = right mode from the first one + left mode from the second one). We shall discuss momentarily which class of distributions is wide enough to reflect real life and narrow enough to give us a chance to benefit from the CLT.

III. Serious uncertainty

In this scenario we have a *definite* fork either for the outcome or for the time to completion, which may influence the critical path or even the outcome of the entire project. Speaking of bimodal (or even multimodal) distributions, let us examine the idea of "forks". Surprisingly, they may pop-up during a perfectly deterministic planning at any stage of any project. The most familiar and developed application is the contingency planning that is used to deal with some disastrous (for the project completion) events. There is a well-developed technique that permits assessing the probability of such events (e.g. major flooding that makes a company's headquarters inaccessible), their negative impact on the planned project, cost of prevention or creation of "plan B", and finally making a decision whether we should create those contingency plans at all—after all, maybe doing nothing or just buying insurance is the best solution. What is more interesting for us is the *casual* forks that represent situations that are out of our control, at least for the purpose of preliminary planning. Let us look at a typical situation—availability of resources. Suppose we have some job that can be completed by one worker in 12 days; another worker can do the job in 6 days; and working together they can complete the job in 4 days. If the estimated probability of getting the second worker on board is 0.5, we have a simple fork: our task will take either 4 or 12 days with 0.5 probability for each case. Depending on specifics of our planning, it may or may not affect the critical/optimal path. The situation may get more complicated if availability of the first worker is, say, 0.5. Then the combined fork will have 4 opportunities with 0.25 probability for each of the 4 different outcomes: completion in 4, 6, 12 days, or non-completion at all.

Future complications: In reality, the situation may become even more complicated, if the workers are available with probabilities p_1 and p_2 (not equal to 0.5), and we have a contingency plan of securing a third worker who can finish the work in 15 days with probability 1 (provided the two original options are not feasible). While this situation is extremely common, it is obvious that it can ruin any meaningful planning, if such uncertainties are not resolved as soon as possible. Now let us throw in one more uncertainty—suppose the numbers of days for completion are, in fact, intervals: 5–7, 10–14 and 12–18. If we know an expected distribution for every worker, then we can create the expected distribution for the entire task, but a) it will be multi-modal, which means that the mean time to completion will be meaningless; and b) the 95% confidence interval for the time to completion will be very wide (in our example between 3.5 and 17 days) and will require careful interpretation.

The last remark is actually very important, and it justifies the use of the analysis of forks in principle. For example, in our original case with 2 workers, it is obvious that we need a contingency plan due to a relatively high probability (25%) of the worst case scenario (non-completion), and a significant increase in time to completion, especially if this increase will interfere with the critical path. In such a case we may consider a better mitigation strategy, e.g. improve probability of securing the best worker, or finding a better replacement than the worst worker. If this part of our planning is really critical, we can apply even more dramatic changes to improve the results before we would even start the project. Needless to say, all this is oversimplification, just to demonstrate the idea.

Now we can see which elements should be added to deterministic planning to make it more realistic. Let us call the *fork*-type uncertainties the *probabilistic* element and *random distribution*-type uncertainties the *stochastic* element. Strictly speaking, as we just demonstrated above, both of these elements are not alien to the classic deterministic planning. Either a classic uncertainty tree or planning of elementary tasks using intervals (instead of numbers) is pretty common.

In addition, we shall assume that we have a well-developed *estimation* technique that works reasonably well in general, both for a set of probabilities for each fork and for the mean and the range of time to completion for relatively simple tasks/subprojects. Another important point is that the resolution of uncertainties is not different in the dynamic planning from handling of errors in estimation in the classic deterministic planning—generally it is re-planning after some non-correctible deviation has occurred.

Now that we have a better understanding of what we are dealing with, let us formulate our task.

After some preliminary work, the first author "modestly" defined *the goal of the project* as the development of a planning tool with the following properties:

- Easy to meaningfully interpret the final assessment.

- Time to completion oriented.

- *Dynamic*—automatically repeated after each completed (or not) subtask.

- Built-in self-correction mechanism—continuous assessment of deviations and early detection of the need of major re-planning.

- Accounting for uncertainties of real life (forks).

- Accounting for stochastic variability (random distributions).

- *Reliability*—this looks a little strange, but it is another side of being realistic and honest.

- Balance between accuracy, precision, and the efforts needed to achieve them; in other words *practicality* and *pragmatism*.

- *Superiority* over existing planning tools with stochastic elements.

The last point looks like an ambitious goal, but the mere fact that the first author was hired to work on the problem is a good indicator that existing planning tools are not always satisfactory. Then the *list of intermediate tasks and problems* was formulated as follows:

- Overview of existing planning tools and methodologies for risk assessment.

- Development of convenient terminology.

- Formulation of a new (presumably better) approach.

- Development of the theoretical base for this approach.

- Formalization of the entire project:

  - structuring, defining the scope;
  - creating a list of pre-requirements and assumptions; and
  - identification of existing methods that could be used (possibly with proper adjustments).

- Division of the project into executable stages (e.g. right now all work is focused on the estimation time to project completion).

- Creation of supportive programming package.

As one can see, many of these steps have already been discussed in this chapter, but mostly informally. We shall now go through the entire list of what was done with the following clarifications:

1. The authors have decided to avoid heavy hard-core math and statistics in the presentation (which essentially serves just to support the ideas that are very simple and are much more useful than the underlying science).

2. The programming package was pretty simple and extremely boring in technical details—any proficient programmer can (after spending some time, of course) recreate it (probably even better than the first author) using clearly formulated underlying ideas.

3. Obviously, two of the above items are outside of this book's scope.

Of course, it was important to have an *overview* of the background of current practices used by professional estimators of (at least in IT) projects.

Note: Unlike in the theory of statistics, in our context by *estimators* we mean the individuals who are tasked with providing quantitative estimates of the project costs, resources or duration. *Estimation* is the process and an *estimate* is the outcome of the estimator's work.

Here is the overview of current practices:

- Classical deterministic planning is completely abandoned, labeled as naïve and amateurish and replaced by the so-called *uncertainty cone*, which is believed to gradually decrease by itself over the life of a project.

- The major idea is that the only viable approach is estimation of each subtask using estimation of the interval of possible outcomes, accompanied by some specific (established in practice) distribution that would take care of all uncertainties associated with a particular task.

- The planning is mostly done based on the *median* of the distribution, even though it is known that such planning is overly optimistic. In order to adjust the results, most estimation tools use some mysterious upgrading coefficients and other tricks.

- Summation of distributions for both sequential and parallel tasks is performed using Monte Carlo techniques.

- Some authors apply special treatment for risks, but, in fact, a common belief is that the only problem is to properly estimate the corresponding distribution.

- The picture of a distribution with a steeper left side and a heavy right tail is viewed as the "law of nature" for the time-to-completion of IT projects—generally, all estimation tools try to adjust all "fair estimates" in this direction (the Pareto family of distributions is frequently used).

Let us see where possible improvements can be made in this setting. First of all, a diminishing value of deterministic planning is simply counterproductive, considering the fact that the planning was actually perfected there. Next, estimation of the time to project completion based on an empirical distribution does not look right—we already have seen that studying built-in uncertainties should provide more valuable insight. Using the median for planning has two major limitations which we will discuss in detail momentarily.

Now, in short: a) it came from aggressive planning and turned out to be indeed overly aggressive; therefore it needs some clever corrections to become realistic again; and b) it complicates simple operations on distributions; e.g. for two sequential independent tasks, the mean is the sum of two means, but for medians you need to do a lot of extra work.

Using Monte Carlo simulations for complicated algorithms is beyond the first author's understanding of statistics—summation of distributions directly using the frequentist approach with pre-defined precision seems to be a much more streamlined approach.

Risk management is a very well developed, integral part of deterministic planning and all attempts to improve it actually boil down to one of the three possibilities: *risk mitigation, contingency plans,* or *full disregard* (based on valid analyses, of course). We will see later that there is a less mystical explanation for the famous curve, but the real "red flag" is an artificial adjustment of the theoretically correct estimates.

Based on these expected improvements it was possible to formulate the *ideology* of the proposed approach:

- Based on *deterministic* planning.

- All systematic factors are separated and included in deterministic planning.

- Random fluctuations around the "true value" and uncertainties related to estimation of this value are handled separately (this is, in essence, separation of probabilistic and stochastic elements).

- Small deviations from true values are considered either a part of random fluctuations (effectively ignored) or a subject of the error correction mechanism (this would happen only if they were detected as errors).

- Big variations of true values are viewed as discrete random variables of "gambling type" when the set of values with corresponding probabilities is the subject for estimation (this is a practical criterion to detect hidden uncertainties inside abnormally wide estimated distributions).

- Factors that are completely out of control are considered to be pure uncertainties; all others are viewed as part of planning and are the subject for risk mitigation.

- Random fluctuations are studied as combinations of independent symmetrical (uniform or normal) relatively tight distributions (this is a critical element of the proposed approach and we will spend considerable time discussing this idea).

- Uncertainty tree with forks being resolved over the project lifetime replaces the currently used uncertainty cone.

- All summation (along and across branches) is done directly with probability density function (PDF) (no Monte Carlo simulations).

The most important element that actually defined the entire ideology was the pragmatic approach. In fact, the choice was made in favor of *simplicity* and *practicality* over complexity of structure and multiple improvements and adjustments (with questionable effect on the quality of planning). The main principle was: simplest solutions are the best choice—we are forced to trade almost anything for simplicity in order to create a viable model. The main questions to be answered before making any change in modeling and the algorithm: Will a more complicated approach improve accuracy of the model at all? If the answer is "yes", the next question is: Will the magnitude of improvement justify extra (potentially impassable) difficulties?

Before we proceed with more formal definitions and a description of the proposed algorithm, let us discuss two issues that can seriously affect the outcome of an open-ended project, and which definitely belong to deterministic planning: *errors in estimation* and the *unforeseen* on the preliminary stage factors. For truly dynamic planning, the key for handling these issues is early detection and prompt and objective evaluation of the impact. A possible impact on the outcome can vary from none to catastrophic (leading to project termination), with anything in between. Statistics can only help in detection of big or systematic estimation errors. Small errors that are within the planned range are undetectable, unless we have repeated estimations that are systematically located at one side of the estimated "true value".

7.3.3   Terminology and definitions

We are now ready to give more or less formal definitions of the key terms.

Uncertainty is an event with multiple possible outcomes that significantly affects our project. The major assumption that makes our planning even theoretically possible: Each outcome has a probability, and our estimation technique permits listing of all such outcomes and assigning the true probability to every outcome. Instead of creating an *estimate* (E), we are going to create multiple estimates $\{E_i\}$. Graphical representation makes the terms *fork*, *branch* and *branching* self-explanatory.

A single uncertainty with (finite) number of outcomes is described as follows (cf. Figure 7.2 for an example of an uncertainty with three outcomes). Suppose $\{A_i\}$ is the full set of mutually exclusive possible outcomes with estimates of interest $\{E_i\}$ and probabilities $\Pr(A_i) = p_i$, $\sum_i p_i = 1$ and $\Pr(A_i \cap A_k) = p_i \delta_{ik}$ ($\delta_{ik} = 1$ if $i = k$ or 0 if $i \neq k$). The simplest uncertainty is a *dichotomy*: event A has a probability p and the complementary event $\bar{A}$ has probability $(1 - p)$.

Uncertainty Tree

Let us consider a scenario when all uncertainties are dichotomies. Suppose we have an all-inclusive set of events $\{A_i\}$ with corresponding probabilities $\{p_i\}(1 \leq i \leq n)$ that address all uncertainties in our project. The space of elementary events has 2^n elements—each element is a vector that has length n and entries 0 or 1. The "1" at the ith entry means that event A_i occurred, and "0" means that A_i did not occur. If all dichotomies are independent events, then the probability of an elementary event in the space of events is just the product of two products $\prod_i p_i$ for all events that occurred and $\prod_j (1 - p_j)$ for all events that did not occur. In the end, we have exactly the same situation which we have seen already for a single uncertainty with the finite number of values: 2^n estimates with 2^n probabilities (cf. Figure 7.3).

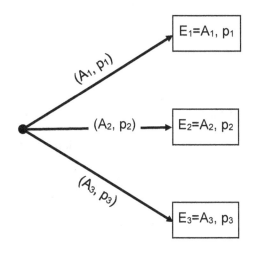

FIGURE 7.2
A single uncertainty with three outcomes

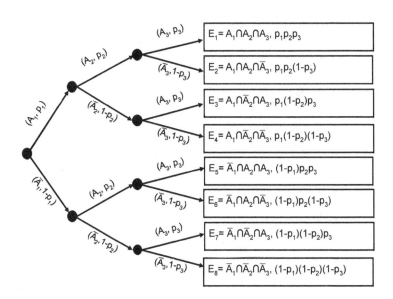

FIGURE 7.3
An uncertainty tree with three dichotomies

In a real project, it is likely that not all branches will be presented, and the conditional probability of an event A_n could depend on the outcomes of the previous set $\{A_i\}$. We shall leave it to the reader to figure out what happens with this uncertainty tree in the case of a real life example with 3 workers from §7.3.2, when A_i is the outcome of the ith worker's participation in the project.

It is almost intuitive that any open-ended project is associated with its uncertainty tree. When this technique is applied to classic deterministic planning, we will end up with a finite number (say, N) of branches, where each of them has an estimated probability and is the subject of separate planning. If we are interested solely in time to completion, we will have at the end the table of outcomes with assigned probabilities.

Despite the fact that we have not accounted yet for the stochastic element of random fluctuations, we can already make some useful preliminary assessments.

We can separate branches when the project is not finished—the branches where it would take unreasonably long to complete the project, and the branches that represent more or less satisfactory completion—and calculate estimated probabilities of, say, 3 main outcomes of interest: $O_1=\{$completed before time $T\}$, $O_2=\{$(can be) completed after time $T\}$, and $O_3=\{$cannot be completed$\}$. We can analyze the branches that produced the undesired outcome in order to better understand the sequences of events that could potentially derail our project and start thinking about possible risk mitigation.

Let us make it clear that the most developed part of deterministic planning—contingency plans for extremely rare catastrophic events—is paradoxically our smallest worry in the overwhelming majority of projects that we are planning.

Moreover, from a practical perspective, we should establish some reasonable threshold for the probability (somewhere in the 1–2% interval), below which we are not including an event in our planning at all. After all, we should understand that intrinsic preciseness of our planning makes working with such events a counterproductive exercise.

The possible mitigation of an ultimate failure can be considered in two directions:

1. Establish the uncertainties that could be (at least partially) influenced, and assess whether the game is worth the candle.

Example: Suppose we are aiming at BLA approval and we want to improve our chances for success (which are, say, 50/50) before submitting it. We detected three factors that, by our evaluation, can significantly improve probability of success (say, to 80/20):

a) Conduct an additional RCT to make our case much stronger (ideally, bulletproof).

b) Conduct additional analyses of the already collected data (that may turn out to be a viable substitution for an RCT).

c) Improve quality of our BLA by investing extra time and money in its preparation.

The list has standard features of real-life decisions: the expected impact, the required time and the cost are going down from a) to c). The most plausible scenario would be, perhaps, skipping a), since it will take a lot of time and money and could be dicey (because the results cannot be guaranteed). A near-optimal decision would be probably to conduct b) and c) in parallel, utilizing all available resources within the company, and using outsourcing as the last resort. Proper planning and execution of c) may be challenging: while clean data, accurate programming, decent medical writing, and overall technical quality of the BLA package will undoubtedly improve our chances for approval, there will be a point of diminishing return-on-investment, and extra delays can turn out to be extremely costly to the company. And let us not forget to allow for extra time and effort (or not, depending on the results) of item b) from the very beginning, because the study results may require a major restructuring of the BLA!

2. Establish the uncertainties that are completely out of our control (these could be enhanced remnants of the already established uncertainties we can have an influence on), and try to move them as close to the initial stages of our project as possible. The idea is two-fold: either cut the losses and terminate the project ASAP, or attempt a major re-planning of the project with more modest, achievable targets.

Example: In a clinical development program, one naturally wants to know (objectively) as much as possible about the drug's efficacy and safety. On paper, phase I and phase II studies are not supposed to bring surprises after a carefully conducted pre-clinical program, and phase III is supposed to be a mere formality to seal the deal. Let us say, usually this is not the case. It may be prudent to not just "play by book", but to "go the extra mile"; e.g. if we have any doubt whether our translation of pre-clinical results gives us full info about what to expect in clinical studies, we should seriously consider conducting an extra pre-clinical trial. If this trial gives answers to our questions, it will pay off in the long run!

Another well-known issue is not paying attention to safety signals in small phase I studies—it is in our best interest to confirm or reject them in phase II, before embarking on a long and expensive phase III program. Another piece of advice comes from our personal experience—we would never know it had we not experienced the nightmarish BLA submission described in Chapter 6. When going into a pivotal trial, we should make sure that we are going to test our drug against the *same* comparator that was used to get promising preliminary efficacy results.

Now, let us look at origins of the uncertainty cone that is actually an unavoidable feature of the ideal deterministic planning after applying the uncertainty tree. If we restrict our planning to successful projects only, we will have the range of time to completion as an estimate from the very beginning. When we are moving along the tree, some uncertainties are resolved (we have passed several forks); we have fewer branches to consider, and therefore the range of possible outcomes gets narrower. After we pass the last fork (all uncertainties resolved), our estimate of time to completion, in theory, boils down to a single number. In reality, we are dealing with some range until the project is fully completed.

Another important issue is the meaning of the words *significant impact* in our definition. This includes, without doubt, any change in planning for the branches that originate from the fork of interest. The question remains open if there are no changes—future planning is absolutely identical. Do we really need to introduce extra complexity, or can we do without it? We will get to this part while discussing estimations of elementary tasks a little later.

The last topic that we want to mention before moving to stochastic estimations is summation of estimates across the branches. In deterministic planning, this is not an issue—we just have to put together a simple table of outcomes with corresponding probabilities. In the complete algorithm that accounts for random deviations from estimates, the outcome for every branch is an estimated distribution of possible times to completion.

7.4   Estimating distribution of time to completion of an open-ended project

For an estimated distribution, one important/critical question is the choice of the presentation of the results. Following the declared principles of simplicity and practicality, we will go with a table (or a graph for visualization) of expected probabilities for the pre-defined unit of time, which means the probability mass function (PMF). All advantages of the PDF are essentially exhausted during the first action of the algorithm—estimation of the distribution of time to completion. In very rare exceptions will any arithmetic operation that we are planning to perform on distributions retain us inside of the class of convenient, well studied (let alone easy to work with) distributions. The PMF tables permit direct manipulations that we need to get the final distribution:

1. Combining several distributions at a fork of the uncertainty tree.

2. Combining two sequential distributions.

3. Combining two parallel distributions.

Another important advantage is that the estimator is not limited to continuous distributions or distributions with a classical pattern, especially with a short range of possible outcomes that behave more similarly to forks than to continuous distributions (e.g. a set of probabilities of 0.2, 0.5 and 0.3 to complete task in 2, 3, and 4 days respectively). In cases when estimation is made in the PDF form, there is no problem to convert it to the PMF and store it as such.

Programming and statistical notes: The just described operation #1 above is the easiest to program, because the resulting PMF is a simple weighted average (according to the fork description) over the range of possible outcomes, which is the union of ranges for each distribution. There is no problem to combine any number of distributions at the fork (it will be a mixture of distributions). Operations #2 and #3 require involvement of the Cartesian product of ranges of outcomes for each distribution; thus they should be performed only using two distributions at a time. (Of course, with SAS PROC SQL one can program any finite number of distributions combined, but it will take more computing time and usually will not work for 4+ distributions with ranges including near 100 points.) The main statistical assumption is independence of two combined distributions. The idea can be best illustrated by one of the oldest examples in probability theory—creation of the PMF table for the sum for rolling two dice simultaneously, where 6×6 table of probabilities of the outcome is converted into the PMF table over the range 2–12. Programming of operation #3 is not any different—it is just done for the maximum instead of the sum, and the range for the PMF will be 1–6. It turned out to be practical to cut both tails of an estimated distribution using the 1% threshold.

Finally, we would like to add a note about choosing the unit for our PMF. To have the possibility to work with continuous estimates, we need to choose a unit much smaller than our unit for the final report. The consequence is such that even the exact (deterministic) number may become a uniform distribution, unless we process it as the deterministic number without variation in all branch calculations.

7.4.1   Surprising results of first test runs of the algorithm

The case of "traditional" planning: Suppose that traditional, deterministic planning for some project gives us 100 days as the estimate. However, the project is complicated enough, and we estimate that its completion time is actually normally distributed around 100 with $\sigma = 10$. Then we introduce a very simple uncertainty tree: there is a 20% probability that we are off in the projection, in which case the completion time is $N(140, 10)$—normally distributed around 140 with the same $\sigma = 10$; and a 5% probability that we are significantly off with the productivity of our team, in which case the completion time is $N(190, 10)$. The distribution of expected time to project

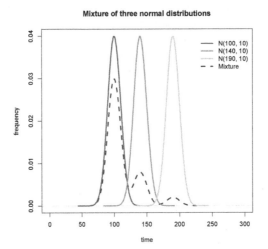

FIGURE 7.4
Distribution of time to complete the project (dashed line) obtained by combining three normal distributions: $N(100, 10)$, $N(140, 10)$, and $N(190, 10)$ at a fork of the uncertainty tree with $p_1 = 0.75$, $p_2 = 0.20$, and $p_3 = 0.05$.

completion is presented in Figure 7.4. This is exactly what we would expect from a 3-modal distribution.

A more realistic estimation: In reality, we probably have a wider spread for the second and third distributions than for the first one (just because longer projects may be also more variable), and we should be ready for a "not so tight" estimation. Figure 7.5 shows what we should expect with σ equal to 10, 15 and 25 (the increase is a little more than proportional to the mean). We still can see the second mode, but the third mode almost disappeared.

A conservative estimation: Now, suppose our estimations are not so tight; say, we multiply all σ values by 2. Figure 7.6 shows a very different distribution! It is clearly unimodal and looks amazingly familiar—this is a "classical" distribution of the time for IT project completion that is referenced in every estimation book as an empirical, very well established and proven observation!

At this point we can make two important intermediate observations. First, the considered example gives us a good idea about the real situation in IT business planning. Second, it adds credibility to the tool that we are developing. Speaking of the model, there are some lessons learned from this first run:

- We do not need any special tricks to create the right shift and raise the right tail of expected/estimated distribution—accounting for several plausible uncertainties from overly aggressive planning will do the job.

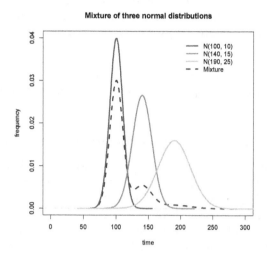

FIGURE 7.5
Distribution of time to complete the project (dashed line) obtained by combining three normal distributions: $N(100, 10)$, $N(140, 15)$, and $N(190, 25)$ at a fork of the uncertainty tree with $p_1 = 0.75$, $p_2 = 0.20$, and $p_3 = 0.05$.

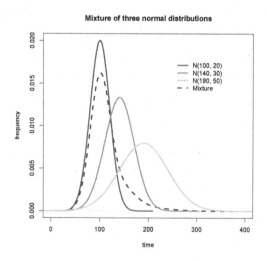

FIGURE 7.6
Distribution of time to complete the project (dashed line) obtained by combining three The normal distributions: $N(100, 20)$, $N(140, 30)$, and $N(190, 50)$ at a fork of the uncertainty tree with $p_1 = 0.75$, $p_2 = 0.20$, and $p_3 = 0.05$.

- Uncertainties that are presented in our planning for a single task will not necessarily lead to a multi-modal distribution, but the presence of at least two modes/peaks definitely means the presence of uncertainty that should be accounted for.

- Even the simplest basic estimation distributions, combined with built-in uncertainties, can create practically all possible distributions.

7.4.2   The nature of estimates for elementary tasks

Before establishing/positioning the previously introduced/described stochastic element into our planning model/algorithm, let us describe some scenarios which should *not* be part of stochastic planning.

The first scenario is when there is no doubt whatsoever in the time to complete the task, and this task belongs to deterministic planning. This is actually a very strong requirement for a task which includes: i) very low probability (below the chosen level of detection) of events/factors that can change the outcome, and ii) very short execution time relative to the used unit of time. There is no need to explain point i), but let us make some notes on the necessity of point ii). When we looked at the example of a task that takes 2–3 days and has possible variations in estimation of hours, we saw two possibilities: either it is just one number from the beginning, or (which is preferable for the purpose of deterministic planning) it could be achieved by assigning a larger number to the estimate (e.g. just 1 day larger). A situation when this estimate impacts the critical path will be accounted for in the second case described below. Of course, if an estimated task has some projected length, say 40–50 days, there is usually no way to know exactly the number of days, and even a 99% confidence interval will have some non-zero width (e.g. 3–5 days).

The second case is actually a small expansion of the first case: the range of possible estimated time is very short, say 5–7 values (not necessarily a full interval; for example, 3, 4, 7, 8, 10, 11, and 12) which makes direct creation of PMF more practical than introducing a stochastic element. This may arise in a situation similar to the example of availability of the 3 workers to complete the task—essentially, uncertainty with the extremely tight estimates for every branch.

There are several major types of estimates for elementary tasks that may lead to incorporation of the *stochastic* component:

1. The estimator is very certain about the time the task will take to complete. (Note that it seems that most people would agree that the majority of tasks are well known and there is no doubt about how long it will take to do them.) However, it is subject to a random deviation from the planned

time point without any reason, or because of a large number of small reasons that are impossible to study. This is obviously the best-case scenario, closest to the deterministic planning. In this situation, we have possible times to complete the task The *normally* distributed around the planned time μ (which is the mean, median and mode of our distribution). The range that the estimator provides truncates the ideal normal distribution according to the detection level adopted by the estimator. We can define this level either: i) by a fraction of outcomes that we decide to not account for (e.g. 5%, 0.5%, etc.) or ii) by the ratio of expected frequencies at the tails to the peak frequency values in the middle of our range (e.g. 0.1, 0.01, etc.) While strategy ii) seems to be much easier to implement, strategy i) has value in explaining the model. For example, the first described situation corresponds to the case when Range=$[\mu - 2\sigma, \mu + 2\sigma]$, and the second one to the case when Range=$[\mu - 3\sigma, \mu + 3\sigma]$. As a result, we have both the mean and the standard deviation for the estimated distribution.

From a practical standpoint, we should ask our estimator several questions to assure that we are dealing with this best-case scenario: generally, we should ask whether: a) there is "the number"; b) there is a symmetry; and c) whether the frequency in the tails is, at least, five times smaller than the frequency in the middle. If the ratio is less than 0.2, then we can regard this estimation as normal; otherwise, for any practical purposes it will behave as a uniform estimation of the second type, which we describe below. Of course, it is naïve to expect our estimator to assess this ratio very accurately; thus the next question after passing the test for normality is whether the ratio is close to 10 or 100. (Of course, we may find later on that we need one more grade (somewhere near 35) that will correspond to Range=$[\mu - 2.5\sigma, \mu + 2.5\sigma]$).

2. The estimator is very certain about the range and has no preference for any specific part of this range to be expected more frequently, or he/she has no knowledge about how the probability density looks inside the range. It seems that the first part of the sentence describes better the known situation than the second part; however, mathematically, they are equivalent and represent a *uniform* distribution.

The most important question about these types of estimates is whether they are practical for a planning purpose or not. Clearly, the answer depends on the expected range. At one extreme, there is no estimation at all—we admit that it could be any number from 0 to ∞ and we have no preference to select any special interval—in this case we simply cannot make any statement about the outcome. At the other extreme, with a very short interval, we do not need any distribution (as discussed earlier). It is obvious that if the range of such an estimate is comparable to the expected length of the entire project, we cannot have any meaningful planning.

3. *Bimodal/multimodal estimates:* Before we discuss asymmetrical estimates in general, let us examine the situation when our estimator knows not just

one true number, but two (or more) numbers with random fluctuations around them. There is no doubt that the only reason for the estimator to see more than one "true value" is uncertainty incorporated into it. The solution is simple: the estimator will create a table of estimated probabilities for a fork and will give a separate estimate for every "true value". This technique will work for several "true intervals" or, after some adjustments, for any combination of "true values" and "true intervals". Simply put, we are separating our task into several really elementary distributions as branches of the fork associated with the presented uncertainty.

4. The most inconvenient estimates are *asymmetrical* estimates. The situation is quite paradoxical. As we can see from the previous descriptions, if, for an *elementary* estimation you know the true number, you have a normal distribution; if you know the true range, you have a uniform distribution. If you know neither the true number nor the true range, then you do not have an estimate at all. Thus, any elementary estimation has to be symmetrical.

There is no question whether asymmetrical estimations exist or not—they do, which means the only one possible explanation—they are not elementary.

The problem is that it is almost mandatory in all existing planning tools to study asymmetrical estimations—if you end up with a symmetrical estimation, people may call you naïve or very lucky. Most accepted methodologies try to model and work with these complicated distributions using Monte Carlo simulations. In the first author's opinion, small asymmetry in elementary estimation can be ignored completely (if present), and for large asymmetries of non-elementary estimations, none of the proposed approaches is satisfactory. The confirmation of this fact of life is the presence of secret formulas that adjust estimates toward bigger numbers in any existing model.

The first author believes that all asymmetrical estimations in reality are either bimodal or multimodal distributions. In order to confirm this hypothesis, let us analyze how such estimations are usually created. (This visualization represents a very typical case when an estimator uses an approach based on expected occurrences of different scenarios.) At the first run, the estimator creates *one true number*, which is, in fact, his/her estimation. The next step is the assessment of random fluctuations around the obtained estimate. A naïve estimator stops here, but a sophisticated one moves on. He/she evaluates all possibilities for the shortest possible time of execution and extends the left tail of the range, by adjusting the possible frequencies up. After that, the same is done with possibilities that would make the time of execution longer. If these two types of adjustment are balanced, then the distribution remains symmetrical, and the only consequence is the increased range. But in real life there are many factors that may delay the project (Murphy's Law) and, more importantly, they have higher likelihood to materialize (the left side is restricted; the right side is not). Thus, in a typical situation the estimator adds (one by one) multiple distributions that represent all uncertainties incorporated into estimation, without even realizing this. In most cases, the result has very little asymmetry that could be either ignored or adjusted for by replacing it with a symmetrical one with the true number moved into the middle of the range (which gives a little more conservative estimate in comparison with the developed true estimate). In these cases, the only problem is an extra complexity of the technique, without any necessity.

In other cases, when the distribution is significantly skewed, the approach simply would not work. The working hypothesis for our model is that in the vast majority of cases, ignoring asymmetry (by replacing an estimate with a little more conservative version) would work, but in extreme cases we would have to separate uncertainties and assess their probabilities. The idea is that, for any practical purpose, any asymmetrical estimation could be represented either by a slightly more conservative version of the first type or by a mixture of several symmetrical distributions of the first and the second types.

Summary:

All estimates for elementary tasks that need a stochastic component can be classified as:

1. Normally distributed with known mean and variance.

2. Uniformly distributed with known range.

3. A combination of two or more distributions of the first two types with estimated weights (in other words, as an uncertainty between two or more estimated distributions with estimated probabilities).

4. Slightly asymmetrical unimodal estimates can be successfully replaced by type 1, and significantly asymmetrical by type 3.

The recommendations to the estimator on how to obtain such estimates could be formulated in simple language. This formulation may take time and some adjustments, but it is generally clear.

7.4.3 Estimation for a single branch

Along a single branch, for estimating the time to completion we may have numbers (degenerate distributions), short range PMF, normal and uniform distributions, and some distributions that are combinations of all above. As we already discussed, our arithmetics is limited to two major operations: combining two (in statistical terms) independent, either *parallel* or *sequential* tasks. Since all our distributions and numbers are stored as PMF tables, two relatively simple macros provided a sufficient toolkit for our purpose.

Combining k parallel tasks with completion times T_i, $i = 1, ..., k$ gives us the total time to completion $T = \max(T_1, ..., T_k)$. Figure 7.7 (two top plots) show the distributions of T for $k = 1, 2, 3, 4, 5$. The top left plot corresponds

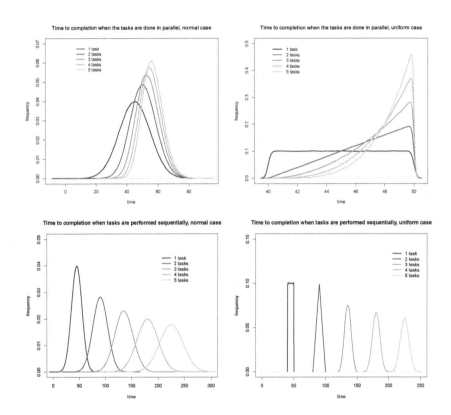

FIGURE 7.7

Distributions of time to project completion combining k parallel tasks (two top plots) and combining k sequential tasks (two bottom plots) ($k = 1, 2, 3, 4, 5$). Top left plot: k tasks are done in parallel, task completion times are independent identical Normal(45, 10). Top right plot: k tasks are done in parallel, task completion times are independent identical Uniform(40, 50). Bottom left plot: k tasks are done sequentially, task completion times are independent identical Normal(45, 10). Bottom right plot: k tasks are done sequentially, task completion times are independent identical Uniform(40, 50).

to the case when T_i's are independent identical normal distributions with mean 45 and standard deviation 10, and for the top right plot the T_i's are independent identical Uniform(40, 50). As k increases, the range essentially stays the same, but the mode of T shifts towards the right end. As expected, starting from 5 parallel identical tasks, >90% of the expected outcomes would be observed in the 10% of the right end of the initial range. Unsurprisingly, the right end of the range becomes the best estimate.

For sequential tasks, the total time to completion T is the sum of the times to complete individual tasks. Figure 7.7 (two bottom plots) shows the distributions of T for $k = 1, 2, 3, 4, 5$. For the bottom left plot, the T_i's are independent normals with mean 45 and standard deviation 10; for the bottom right plot the T_i's are independent Uniform(40, 50). As k increases, the distribution of T is shifted to the right. In the uniform case (Figure 7.7, bottom right plot), the normal approximation becomes visible starting with $k = 3$. This actually supports the idea that if we have 3 or more symmetrically distributed independent tasks in our subproject, we can estimate the entire subproject as normally distributed, even if we cannot do better that using uniform estimates for individual parts. Let us acknowledge that the possibility to plan realistically using intervals solely is confirmed.

Let us now see what we can expect from estimation of a single branch. Most likely, we will have a distribution of the classical type—bell-shaped, right-shifted and right-skewed. Judging by our test runs, there are good chances to have it significantly tighter than the classical one (planning by intervals)—with shorter confidence intervals, and here is why. In interval planning, the range is roughly proportional to the number of combined tasks (for the PMF summation—due to the CLT—to $\sqrt{k}$). The addition of sequential independent tasks with one clear peak indeed tightens the distribution, irrespectively of the shape of the added distributions (due to the CLT). Presence of parallel tasks tightens the distribution even more and guarantees the right shift. The distributions that have built-in uncertainties will, as a rule, be shifted to the right as well, and supply a heavy right tail.

An interesting question is which uncertainties belong to elementary estimation and which to the entire uncertainty tree for the entire project (there is no need to separate uncertainties inside one branch, but outside of elementary estimations). If we want to enjoy advantages of statistical independence, the only uncertainties incorporated into elementary estimation should be clearly related to this particular task, and not influence the others. In addition, such an approach simplifies analyses of the final estimation.

The independence of estimates, in general, should take care of multi-modal elementary estimations and leave us with one distinguishable true value for the entire branch. The uncertainty cone inside a single branch will be present, mostly due to uncertainties that are incorporated into elementary estimations.

What will the final estimation (after summing all branches according to the project uncertainty tree) look like? The answer is "It depends." The number of branches is impossible to predict. The only thing that we can guarantee is that every fork will add at least one branch for a dichotomy and a greater number for a longer list of the outcomes of interest. For a set of n fully independent uncertainties, we will have $\prod_{i=1}^{n} k_i = k_1 \times k_2 \times \cdots \times k_n$ branches, where k_i is the number of outcomes for the ith uncertainty. In reality, dependencies inside the set will lead to a much smaller number. While combining distributions from each branch, we have to keep in mind that a significant part of the branches may correspond to a non-finished project, and thus are not subject to combination. In the worst case scenario, we will get a multi-modal distribution over a long range with a mode for every accounted branch. In reality, we will have a smaller number for two reasons: 1) there will be plenty of branches with such small probabilities that will not create a mode; and 2) some modes (assuming they are randomly distributed over the studied range) will coincide. This is not necessarily a good thing—the complete absence of well-shaped peaks may lead either to almost uniform distribution over a long range, or to a combination of almost uniform distributions that are spread over the same range.

7.4.4   How to analyze the results?

One of the most curious/mysterious features of current estimation practice is the judgment of quality of a proposed model/algorithm/technique/tool by its ability to provide "converged" estimation with a 95% confidence interval. It defies common sense.

If an algorithm is correct, which means more or less accurate and reliable, the result should be solely the function of the project's shape and its readiness for a meaningful estimation. If we have too many uncertainties associated with the project, the estimation must be vague (if we are looking for a realistic estimation, of course). Bad/non-conclusive estimation should be viewed as a tool to understand the reasons, and for creation of a reasonable mitigation strategy, or, at a minimum, as a signal to keep our expectations low about the project outcome. There is no doubt that even "converged" estimations may have bad elements that should be carefully examined.

The final distribution can provide very important supplemental information on what can go wrong, and possibly add additional branches that will not be finished due to a prolonged time of project execution. Another necessity is to locate events that lead just to prolonged execution of the project, without threatening its full failure. After detection of these adverse uncertainties, a mitigation strategy can be developed using the uncertainty tree, which is not only much easier to work with, but also is a map of the project's potential troubles.

In any project estimation, there will be a relatively small number of branches that have significant probabilities to be accounted for—they represent our mainstream planning and should be analyzed similarly to the entire project with creation of a mitigation strategy when it makes sense. In most cases, for a single branch, mitigation just means securing in advance availability of the work force and other resources. The best strategy would be re-planning of the project according to the findings and the created mitigation strategy; staying informed of the remaining uncertainty tree; and re-evaluation of the planning upon passage of any significant fork.

7.5 Summary

In this chapter we have attempted to provide answers to the following questions:

1. Can we make a realistic estimation of the time to completion for open-ended projects using simple tools, previous achievements in deterministic planning, a transparent model, basic science, and common sense (i.e. pragmatic approach)?

2. Will this estimation be of any practical use?

3. Will it be possible to sell?

The answers are:

1. Definitely YES. The idea of the presented algorithm is, in simple words, a Cartesian product of 3 terms: 1) *deterministic planning* with built-in dynamic errors handling; 2) the *probabilistic element* of an uncertainty tree associated with the project; and c) the *stochastic element* of purely random fluctuations around true values.

2. To some extent. It will not solve all the problems associated with the planned project, but can serve: i) as an eye opener; ii) as a basis for mitigation strategy and probable re-planning; and iii) to support dynamic (re-)planning of the project along its progress.

3. As for now, definitely NO. *Realistic* planning is not competitive, not only in contract bidding, but in principle.

Appendix A: Relativistic and Probabilistic Functions

The *relativistic* function (summing velocities in relativity theory when they are expressed as a fraction of speed of light) is given by

$$F_R(x, y) = (x + y)/(1 + xy).$$

The *probabilistic* function (probability of event $A \bigcup B$, where $\Pr(A) = x$, $\Pr(B) = y$, and $\Pr(A \bigcap B) = xy$) is given by

$$F_P(x, y) = x + y - xy.$$

3D plots for both $F_R(x, y)$ and $F_P(x, y)$, as well as for the difference $F_R(x, y) - F_P(x, y)$ for $x, y \in [0, 1]$ are provided in Figure A.1.

Let us show that $\max(x, y) \leq F_P(x, y) \leq F_R(x, y) \leq 1$ for $x, y \in [0, 1]$. We have $1 - F_P(x, y) = (1 - x)(1 - y)$ and $1 - F_R(x, y) = (1 - x)(1 - y)/(1 + xy)$. Then the difference $d(x, y) = F_R(x, y) - F_P(x, y) = xy(1 - x)(1 - y)/(1 + xy)$. Clearly, $d(x, y)$ is a non-negative, symmetric function of two variables defined on Cartesian product $[0, 1] \times [0, 1]$ that equals 0 on the boundaries and has the single maximum (approximately 0.05), close to the point $x = y = 0.5$ (cf. Figure A.1, bottom plot).

To investigate more carefully the behavior of $F_R(x, y)$, $F_P(x, y)$, and $d(x, y) = F_R(x, y) - F_P(x, y)$, we looked at some cuts by the planes that are parallel to the z-axis. The biggest difference is, of course, in the cut by the plane $x = y$ (top left plot of Figure A.2).

The most interesting features of these plots are observed when one of the two scores is fixed (e.g. we fix x at some constant value $c \in (0, 1)$ and look at the functions $z = F_R(c, y)$, $z = F_P(c, y)$, and $z = F_R(c, y) - F_P(c, y)$). Three plots in Figure A.2 (top right, bottom left, and bottom right) give visualizations for the cases when $x = 0.32$, $x = 0.53$, and $x = 0.84$, respectively. For the probabilistic function (dashed lines in the plots), the dependency is linear. By contrast, for the relativistic function (solid lines in the plots), the relationships are concave. This is the reason why we chose the relativistic formula—it penalizes more for multiple AEs. The occurrence of two (even independent) AEs may create a synergistic effect on the outcome. Of course, this penalty diminishes when we are nearing an ultimately bad score of 1.

In general, a concave family of functions would put more penalty on AEs with lower scores, and probably, would work better to separate outcomes

189

in trials when the majority of patients have very low scores. By contrast, a convex family of curves would work better in populations with high scores and it would be less sensitive in populations with low scores. Theoretically it is possible to create curves that would go from convex to concave for different intervals, but besides being too complex, such summation functions would, on average, behave similarly to $z = F_P(x, y)$. While it is possible to propose a clear algorithm for creation of different summation functions with specific properties, generally all of them would give very similar results. For instance, as we have seen from Figure A.1, the difference in assessments based on the relativistic and the probabilistic formulas is negligible.

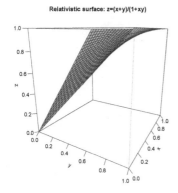
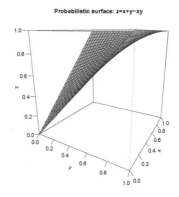

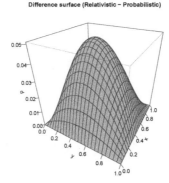

FIGURE A.1

Top left plot: the relativistic surface $z = (x + y)/(1 + xy)$. Top right plot: the probabilistic surface $z = x + y - xy$. Bottom plot: the difference (relativistic minus probabilistic), expressed as $z = xy(1 - x)(1 - y)/(1 + xy)$.

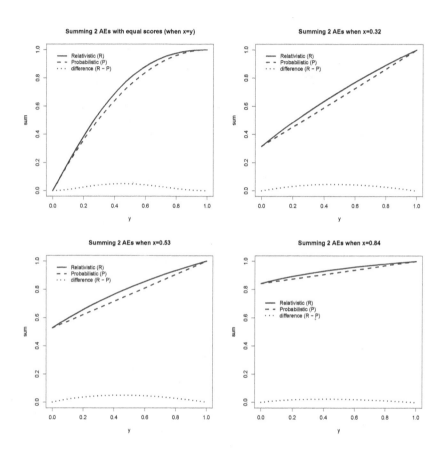

FIGURE A.2
Behavior of the relativistic (R), the probabilistic (P), and the difference (R–P)
functions for some cuts by the planes that are parallel to the z-axis. Top left
plot: the cut by the plane $x = y$. Top right plot: the cut by the plane $x = 0.32$.
Bottom left plot: the cut by the plane $x = 0.53$. Bottom right plot: the cut by
the plane $x = 0.84$.

Appendix B: Manual for Successful Crusade in Defense of Patients' Rights

Pitman, A. N. and Zhiburt, E. B.

Incorrectly conducted meta-analysis could serve as a foundation for sensational revelations about harm caused by a particular drug or a group of drugs during their history of usage. The authors have created an ironic Manual for a running modern "scientifically sound" crusade. On a serious note, minimization of random and systematic errors in a chosen methodology is a prerequisite for achieving trustworthy results while performing meta-analyses.

Key words: *meta-analysis, medicines, pseudo-sensations.*

Occasionally, in the field of clinical transfusion, we witness very heated debates that follow a typical scenario:

a) Publication of a newly found adverse effect of a drug.

b) Extremely emotional discussion in press.

c) Publication of serious deficiencies in assembling of the initial statement.

d) Realization that there is no reason for concern.

This was the case with findings about increased mortality due to infusions of albumin[1-3], increased risk of myocardial infarction due to infusions of hemoglobin-based oxygen carriers,[4,5] increased risk of non-Hodgkin lymphomas after hematologic transfusions.[6,7]

To people who are willing to organize such campaigns, we propose a simple algorithm of actions based on meta-analysis, which has been proven to be very useful in such situations. The recipe is simple:

• Declare that you are going to use meta-analysis (a widely accepted technique in the scientific community) to uncover some truth about a group of drugs.

• Assure the audience that you are going to follow the existing guidelines (especially if you are planning to ignore them completely).

- Create a criterion for pooling different sources of data (usually "class effect" for a group of drugs).

- Collect as much data as possible, to increase the sample size and to make the meta-analysis all-inclusive, thereby creating a foundation for generalization of the results.

- Select an outcome which would reveal statistical significance (drug vs. control) in the combined analysis dataset.

- If necessary, try several search criteria, to obtain the desired outcome.

- Replace the question about scientific validity of combining various (heterogeneous) datasets by statistical test of homogeneity (e.g. the Breslow-Day test), and

  - if positive, declare combination of data valid;
  - if negative, declare combination valid after changing the model from the fixed-effect to a random-effect.

- Choose a hypothesis that "has been proven" by the observed outcome (surprisingly, it is usually again the "class effect").

- Make sure to present the chosen hypothesis as the only (known to the scientific community) possible explanation of the observed outcome.

- Further "prove" the class effect and validity of combination of data by running an exotic subgroup analysis, while ignoring legitimate subgroups related to the chosen hypothesis and the real heterogeneity in clinical settings.

- Declare revelation of the "hidden truth".

- Make a negative and, at the same time, convenient for citation conclusion.

- List all limitations and state clearly that they by no means influenced the final conclusion, while referring to the results of subgroup analyses.

- Generalize the conclusion as much as it is possible.

- Make some banning recommendations based on this generalized conclusion.

- Make an accusation that this has not been uncovered earlier.

- Make the regulatory body responsible for the situation.

- Launch a public campaign for immediate compliance with the proposed recommendation around the world.

The beauty of this strategy is that it is almost independent from the chosen drug (or the group of drugs), clinical settings, and the actual collected data. The only requirements are the big volume of data and the presence of some safety signals.

If authors have no personal preferences while comparing two different groups of drugs (preferably with distinctively different purposes), the situation gets even better. They can make a final choice of a "killer drug" after obtaining statistical significance, which is almost guaranteed with a sufficiently large sample size. Since there is no need for providing explanation of underlying mechanisms responsible for the results, chances for success are doubled. For instance, in the "Colloids versus Crystalloids" discussion, the "killer drugs" can be either colloids[8] or crystalloids.[9]

References:

[1] Cochrane Injuries Group Albumin Reviewers. Human albumin administration in critically ill patients: systematic review of randomised controlled trials // BMJ.- 1998.- Vol. 317, No.7153.- pp.235–240.

[2] Higgins J.P.T., Thompson S.G., Deeks J.J., Altman D.G. Measuring inconsistency in meta-analyses // BMJ.- 2003.- Vol. 327, No.7414.- pp.557–560.

[3] Zhiburt E.B. To kill or to research? (On the end of albumin and meta-analysis of the common sense) // Medical Paper.- No.29.- 04/20/2005.

[4] Natanson C., Kern S.J., Lurie P., Banks S.M., Wolfe S.M. Cell-free hemoglobin-based blood substitutes and risk of myocardial infarction and death: a meta-analysis // JAMA.- 2008.- Vol.299, No.19.- pp.2304–2312.

[5] Greenburg A.G., Pitman A., Pearce B., Kim H.W. Clinical contextualization and the assessment of adverse events in HBOC trials // Artif Cells Blood Substit Immobil Biotechnol.- 2008.- Vol.36, No.6.- pp.477–486.

[6] Cerhan J.R. Transfusion and NHL risk: a meta-answer? // Blood.- 2010.- Vol.116, No.16.- pp.2863–2864.

[7] Castillo J.J., Dalia S., Pascual S. The association between red blood cell transfusions and the development of non-Hodgkin lymphoma: a meta-analysis of observational studies // Blood.- 2010.- Vol.116, No.16.- pp.2897–2907.

[8] Bisonni R.S., Holtgrave D.R., Lawler F., Marley D.S. Colloids versus crystalloids in fluid resuscitation: an analysis of randomized controlled trials // Journal of Family Practice.- 1991.- Vol.32, No.4.- pp.387–390.

[9] Velanovich V. Crystalloid versus colloid fluid resuscitation: a meta-analysis of mortality // Surgery.- 1989.- Vol.105, No.1.- pp.65–71.

Afterword

It all started when two of the authors (AP and LBP) had peacefully worked for a company (Holy Grail company, for simplicity) that developed hemoglobin-based oxygen-carrier (blood-substitute). Their common target was analysis of results of the already finished program that culminated in the years 1998–2000 with the pivotal trial (following 15 or even 20 years of pre-clinical and early clinical development). For a while (2001–2003) both worked in parallel, barely knowing each other. Their close collaboration started circa 2003 when LBP was the head of pharmacology and AP was the manager of statistical programming.

In 2008, when the first incarnation of Holy Grail went belly up, AP and LBP were both left up with their valuable heritage and with the necessity to continue to feed their families. While generous New England jobless benefits (about 15% of their salaries at termination) lasted, the two aforementioned tried to capitalize on their developed skills and their achieved understanding of drug development problems, combined with a clear vision of how to avoid them altogether in the future. The intermediate results that brought great intellectual satisfaction without any monetary reward included creation of a website devoted to safety outcome assessment that brought wisdom to anyone who wanted to learn, a workshop in the FDA Translational Science Office and uncountable promising communications with interested parties.

After one year of struggling, LBP went the traditional way and landed in a very respectful job with a hard-working and well-paid consultancy service specializing in pre-clinical drug development (which he successfully keeps up to date), while AP journeyed into a weird unknown territory exploring his unique qualification as a pragmatic planner of unachievable goals. The half-paid off contract with an IT company (described in details in Chapter 7) permitted a year-long survival and, surprisingly, resulted in one more website that promoted a high level understanding of "properly designed and well-executed" clinical development programs. While these websites were generally well-received by a small group of individuals who were interested enough to get through this extremely useful information (free of charge), they have eventually faded out without any visible effect on the drug industry.

The unexpected turn of events occurred when Holy Grail was revived (with the same CEO and under a different name, of course) and invited the adventurer (AP) back. The well-settled LBP was not interested in future excitement. For AP it was an incredible opportunity to apply newly acquired knowledge, despite known rifts between the new Owner and the old (incredibly devoted to ideas) CEO. In spite of incredibly successful efforts to restore the drug's image in the West, the new Owner lost ground on making huge quick money in the East, blamed the predecessors and essentially stopped financing. As a result, in the beginning of 2013 the senior management of the company was let go and a few months later the adventurer (AP) got his severance package as well.

While the senior management was busy getting Holy Grail back for the third birth, AP made his attempt to get a suitable job in a very big CRO, using recommendations of well-positioned friends acquired during the years in Holy Grail. The level of recommendations got him to the proposal: "Make self-presentation to our decision makers." Looking back, it was the conception day of the current book.

The approach that the adventurer (AP) took to create the presentation was, gently speaking, unconventional. The expectation was that the presenter would describe some statistical tricks for fashionable and, probably, unnecessary services—according to the position description as "serving small pharma needs". However, a mismatch between expectation and reality occurred. The adventurer (AP) was a "small pharma" person who had a long experience of interaction and collaboration with big CROs and had accumulated some valuable knowledge on what the small pharma *truly* needs (which did not match the big CRO intentions). Of course, this type of presentation called for some self-promotion—a description of what the presenter had done and what he would bring to the table. The main goals were modestly set to define what the small pharma *truly* needs, which immediately led to *pragmatic problem solving on multi-dimensional levels that requires high coordination by a trusted subject matter expert.* The main idea was to capitalize on a well-known psychological conundrum *"No prophet is accepted in his hometown."* Indeed, you have to earn this position from the inside by putting in many years of being always right and you have it for free if you are an external consultant.

The entire presentation was designed to show how the adventurer's hard-earned knowledge, experience and skills (even when they sometimes seemed to be unrelated to the current job description) made the presenter the best candidate for the open position. Specifically, there were two major concerns to be addressed:

- In the land of statisticians, a mathematician has to prove that he is relevant.

- The relevant experience is not limited to your work experience. If you work creatively in some field using a well-developed system of principles, you can successfully apply them in other fields as well.

While this looked very plausible and noble on paper, it did not work career-wise. Frankly, all insiders warned the adventurer (AP) that it would not work. Despite this, they provided valuable help to shape the presentation, and the biggest help, as expected, came from the long-time Holy Grail collaborator

(LBP). The best feedback that the adventurer cherished most was from an outsider with big pharma experience: "...They may call 911 for medical help or make an offer for a position two levels higher than discussed. It is 50/50—can go either way." In fact, it ended in a decent middle—everybody was entertained, but no job offer followed.

After a year of licking wounds, the adventurer (AP) accepted a job offer from the third incarnation of Holy Grail, and this weird presentation was almost forgotten. But Fate decided otherwise: while the adventurer was visiting a designated bar in Boston with fellow statisticians during an annual Joint Statistical Meetings, there was a "blast from the Past" (OS)—a former student (back to Soviet Union teaching experience, after 25 years without any communications), who became a respectable Ph.D. statistician on his own. When old relationships were restored, the first thing that was discussed was, of course, drug development, and this presentation was given a fresh look, without any intentions. A couple of months later there was a sudden proposal: "If we add a little water, it could become a decent book." The adventurer (AP) and the old collaborator (LBP) both agreed and then all three of us decided to write this book. The idea was to add (instead of water) some specifics under an umbrella of "pragmatic problem solving skills in drug development", including along the way some "purely biostatistics stuff" that was completely disregarded in the original presentation. It was quite a long journey (from October 2015 to December 2018) to write this book. The authors do hope that the readers will get some new insights into the role of the biostatistician in clinical drug development while sharing our excitement for this profession.

Final Remark

"Tilting at windmills hurts you more than the windmills."

Robert A. Heinlein "The Notebooks of Lazarus Long"

As stated in the Preface, the authors did not address "soft" skills that in many cases may lead to the problem solving algorithm presented in the figure below. We can say that we tried to cover cases when this "bulletproof" algorithm failed, leaving us in $\boxed{\text{TROUBLE}}$. So, if you have found yourself in $\boxed{\text{TROUBLE}}$, this book may help to develop a better strategy, *especially for those who care.* Of course, there is no guarantee...

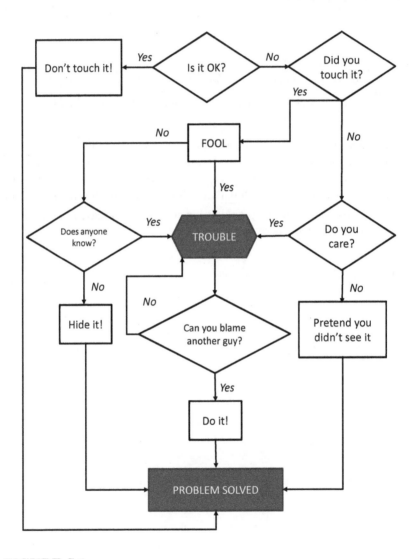

FIGURE S.1
Problem solving algorithm (a joke, of course)

Bibliography

Bibliography

[1] C. J. Adcock. The choice of sample size and the method of maximum expected utility—comments on the paper by Lindley. *The Statistician*, 46(2):155–162, 1997.

[2] V. Amrhein, S. Greenland, and B. McShane. Retire statistical significance. *Nature*, 567:305–307, 2019.

[3] Z. Antonijevic. *Optimization of Pharmaceutical R&D Programs and Portfolios. Design and Investment Strategy.* Springer International Publishing Switzerland, 2015.

[4] D. Ashby. Bayesian statistics in medicine: A 25 year review. *Statistics in Medicine*, 25:3589–3631, 2006.

[5] J. Babb, A. Rogatko, and S. Zacks. Cancer phase I clinical trials: efficient dose escalation with overdose control. *Statistics in Medicine*, 17:1103–1120, 1998.

[6] P. Bauer, F. Bretz, V. Dragalin, F. Koenig, and G. Wassmer. Twenty-five years of confirmatory adaptive designs: opportunities and pitfalls. *Statistics in Medicine*, 35(3):325–347, 2016.

[7] N. Benda, M. Branson, W. Maurer, and T. Friede. Aspects of modernizing drug development using clinical scenario planning and evaluation. *Drug Information Journal*, 44:299–315, 2010.

[8] V. W. Berger. *Selection Bias and Covariate Imbalances in Randomized Clinical Trials.* Wiley, Chichester, West Sussex, England, 2005.

[9] V. W. Berger. *Randomization, Masking, and Allocation Concealment.* Chapman & Hall/CRC Press, Boca Raton, FL, 2018.

[10] S. W. Bergman and J. C. Gittins. *Statistical Methods for Pharmaceutical Research Planning.* Marcel Dekker Inc., New York and Basel, 1985.

[11] D. A. Berry. The brave new world of clinical cancer research: adaptive biomarker-driven trials integrating clinical practice with clinical research. *Molecular Oncology*, 9:951–959, 2015.

[12] T. J. Boardman, G. J. Hahn, W. J. Hill, R. R. Hocking, W. G. Hunter, W. H. Lawton, L. Ott, R. D. Snee, and W. E. Strawderman. Preparing statisticians for careers in industry: Report of the ASA section on statistical education committee on training of statisticians for industry. *The American Statistician*, 34(2):65–75, 1980.

[13] B. Booth and R. Zemmel. Prospects for productivity. *Nature Reviews Drug Discovery*, 3:451–457, 2004.

[14] B. Bornkamp, F. Bretz, A. Dmitrienko, G. Enas, B. Gaydos, C. H. Hsu, F. König, M. Krams, Q. Liu, B. Neuenschwander, T. Parke, J. Pinheiro, A. Roy, R. Sax, and F. Shen. Bayesian adaptive design for targeted therapy development in lung cancer—a step towards personalized medicine. *Journal of Biopharmaceutical Statistics*, 17(6):965–995, 2007.

[15] F. Bretz, P. Gallo, and W. Maurer. Adaptive designs: The Swiss Army Knife among clinical trial designs? *Clinical Trials*, 14(5):417–424, 2017.

[16] F. Bretz, J. Hsu, J. Pinheiro, and Y. Liu. Dose finding – a chellenge in statistics. *Biometrical Journal*, 50(4):480–504, 2009.

[17] F. Bretz, J. Pinheiro, and M. Branson. Combining multiple comparisons and modeling techniques in dose-response studies. *Biometrics*, 61:738–748, 2005.

[18] C. F. Burman and S. Senn. Examples of option values in drug development. *Pharmaceutical Statistics*, 2:113–125, 2003.

[19] J. Chen, J. Heyse, and T. L. Lai. *Medical Product Safety Evaluation: Biological Models and Statistical Methods*. Chapman & Hall/CRC Press, Boca Raton, FL, 2018.

[20] S. C. Chow. Good statistics practice in the drug development and regulatory approval process. *Drug Information Journal*, 31:1157–1166, 1997.

[21] S. C. Chow, J. Shao, H. Wang, and Y. Lokhnygina. *Sample Size Calculations in Clinical Research, 3rd edition*. Chapman & Hall/CRC Press, Boca Raton, FL, 2017.

[22] C. Chuang-Stein, R. Bain, M. Branson, C. Burton, C. Hoseyni, F. Rockhold, S. Ruberg, and J. Zhang. Statisticians in the pharmaceutical industry: the 21st century. *Statistics in Biopharmaceutical Research*, 2(2):145–152, 2010.

[23] C. Chuang-Stein and S. Kirby. *Quantitative Decisions in Drug Development*. Springer International Publishing AG, 2017.

[24] R. H. Coase. How should economists choose? *American Enterprise Institute, Washington DC*, 1982.

[25] S. Day. Changing times in pharmaceutical statistics: 1980-2000. *Pharmaceutical Statistics*, 1:9–16, 2002.

[26] S. Day. Changing times in pharmaceutical statistics: 2000-2020. *Pharmaceutical Statistics*, 1:75–82, 2002.

[27] D. L. DeMets. Current development in clinical trials: issues old and new. *Statistics in Medicine*, 31:2944–2954, 2012.

[28] V. DePuy. SDTM What? ADaM Who? A Programmer's Introduction to CDISC. 2014.

[29] A. Dmitrienko and E. Pulkstenis. *Clinical Trial Optimization Using R.* Chapman & Hall/CRC Press, Boca Raton, FL, 2017.

[30] A. Dmitrienko, A. C. Tamhane, and F. Bretz. *Multiple Testing Problems in Pharmaceutical Statistics*. Chapman & Hall/CRC Press, Boca Raton, FL, 2009.

[31] V. Dragalin, B. Bornkamp, F. Bretz, F. Miller, S. K. Padnamabhan, N. Patel, I. Perevozskaya, J. Pinheiro, and J. R. Smith. A simulation study to compare new adaptive dose-ranging designs. *Statistics in Biopharmaceutical Research*, 2(4):482–512, 2010.

[32] A. Earnest. *Essentials of a Successful Biostatistical Collaboration*. Chapman & Hall/CRC Press, Boca Raton, FL, 2017.

[33] S. S. Ellenberg, T. R. Fleming, and D. DeMets. *Data Monitoring Committees in Clinical Trials: A Practical Perspective*. Wiley, Chichester, West Sussex, England, 2002.

[34] E. I. Ette and P. J. Williams. *Pharmacometrics: The Science of Quantitative Pharmacology*. Wiley, New York, 2007.

[35] European Medicines Agency (EMA). *Guideline on Multiplicity Issues in Clinical Trials (Draft)*. 15 December 2016.

[36] European Medicines Agency (EMA). *Reflection Paper on Methodological Issues in Confirmatory Clinical Trials Planned with an Adaptive Design*. 18 October 2007.

[37] European Medicines Agency (EMA). *Qualification Opinion of MCP-Mod as an Efficient Statistical Methodology for Model-based Design and Analysis of Phase II Dose Finding Sudies under Model Uncertainty*. 23 January 2014.

[38] S. Evans and N. Ting. *Fundamental Concepts for New Clinical Trialists*. Chapman & Hall/CRC Press, Boca Raton, FL, 2016.

[39] R. Fisch, I. Jones, J. Jones, J. Kerman, G. K. Rosenkranz, and H. Schmidli. Bayesian design of proof-of-concept trials. *Therapeutic Innovation & Regulatory Science*, 49(1):155–162, 2015.

[40] Food and Drug Administration (FDA). *Expansion Cohorts: Use in First-in-Human Clinical Trials to Expedite Development of Oncology Drugs and Biologics: Guidance for Industry.* August 2018.

[41] Food and Drug Administration (FDA). *Multiple Endpoints in Clinical Trials: Guidance for Industry (Draft).* January 2017.

[42] Food and Drug Administration (FDA). *Adaptive Designs for Medical Device Clinical Studies: Guidance for Industry and Food and Drug Administration Staff.* July 27, 2016.

[43] Food and Drug Administration (FDA). *Innovation or Stagnation: Challenge and Opportunity on the Critical Path to the New Medical Products.* March 2004.

[44] Food and Drug Administration (FDA). *Innovation or Stagnation: Critical Path Opportunities Report.* March 2006.

[45] Food and Drug Administration (FDA). *Adaptive Designs for Clinical Trials of Drugs and Biologics: Guidance for Industry (Draft).* September 2018.

[46] Food and Drug Administration (FDA). *Master Protocols: Efficient Clinical Trial Design Strategies to Expedite Development of Oncology Drugs and Biologics: Guidance for Industry (Draft).* September 2018.

[47] N. L. Geller. Statistics: An all-encompassing discipline. *Journal of the American Statistical Association*, 106:1225–1229, 2011.

[48] C. Gerlinger, L. Edler, T. Friede, M. Kieser, C. T. Nakas, M. Schumacher, J. Seldrup, and N. Victor. Considerations on what constitutest a 'qualified statistician' in regulatory guidelines. *Statistics in Medicine*, 31:1303–1305, 2012.

[49] S. N. Goodman. Towards evidence-based medical statistics. 1: The P-value fallacy. *Annals of Internal Medicine*, 130:995–1004, 1999.

[50] S. N. Goodman. A dirty dozen: twelve P-value misconceptions. *Seminars in Hematology*, 45:135–140, 2008.

[51] S. Greenland, S. Senn, K. J. Rothman, J. B. Carlin, C. Poole, S. N. Goodman, and D. G. Altman. Statistical tests, P values, confidence intervals, and power: A guide to misinterpretations. *European Journal of Epidemiology*, 31:337–350, 2016.

[52] A. P. Grieve. Do statisticians count? A personal view. *Pharmaceutical Statistics*, 1:35–43, 2002.

[53] A. P. Grieve. The professionalization of the 'shoe clerk'. *Journal of the Royal Statistical Society Series A*, 168:1–16, 2005.

[54] A. P. Grieve. Idle thoughts of a 'well-calibrated' Bayesian in clinical drug development. *Pharmaceutical Statistics*, 15:96–108, 2016.

[55] I. Hatfield, A. Allison, L. Flight, S. A. Julious, and M. Dimairo. Adaptive designs undertaken in clinical research: a review of registered clinical trials. *Trials*, 17:150, 2016.

[56] A. B. Hill. The environment and disease: association or causation? *Proceedings of the Royal Society of Medicine*, 58(5):295–300, 1965.

[57] E. B. Holmgren. *Theory of Drug Development*. Chapman & Hall/CRC Press, Boca Raton, FL, 2013.

[58] International Conference on Harmonisation (ICH). *E9(R1): Estimands and Sensitivity Analysis in Clinical Trials*. 16 June 2017.

[59] International Conference on Harmonisation (ICH). *E8: General Considerations for Clinical Trials*. 17 July 1997.

[60] International Conference on Harmonisation (ICH). *E10: Choice of Control Group and Related Issues in Clinical Trials*. January 2001.

[61] International Conference on Harmonisation (ICH). *E9: Statistical Principles for Clinical Trials*. September 1998.

[62] T. Jaki, A. Gordon, P. Forster, L. Bijnens, B. Bornkamp, W. Brannath, R. Fontana, M. Gasparini, L. V. Hampson, T. Jacobs, B. Jones, X. Paoletti, M. Posch, A. Titman, R. Vonk, and F. Koenig. A proposal for a new PhD level curriculum on quantitative methods for drug development. *Pharmaceutical Statistics*, 17:593–606, 2018.

[63] C. Jennison and B. W. Turnbull. *Group Sequential Methods with Applications to Clinical Trials*. Chapman & Hall/CRC Press, Boca Raton, FL, 2000.

[64] S. A. Julious. *Sample Sizes for Clinical Trials*. Chapman & Hall/CRC Press, Boca Raton, FL, 2009.

[65] E. S. Kim, R. S. Herbst, I. I. Wistuba, J. J. Lee, G. R. Jr. Blumenschein, A. Tsao, D. J. Stewart, M. E. Hicks, J. Jr. Erasmus, S. Gupta, C. M. Alden, S. Liu, X. Tang, F. R. Khuri, H. T. Tran, B. E. Johnson, J. V. Heymach, L. Mao, F. Fossella, M. S. Kies, V. Papadimitrakopoulou, S. E. Davis, S. M. Lippman, and W. K. Hong. The BATTLE trial: personalizing therapy for lung cancer. *Cancer Discovery*, 1(1):44–53, 2011.

[66] I. Kola and J. Landis. Can the pharmaceutical industry reduce attrition rates? *Nature Reviews Drug Discovery*, 3:711–715, 2004.

[67] K. G. Kowalski. My career as a pharmacometrician and commentary on the overlap between statistics and pharmacometrics in drug development. *Statistics in Biopharmaceutical Research*, 7(2):148–159, 2015.

[68] D. Lakens, F. G. Adolfi, C. J. Albers, and et al. Justify your alpha. *Nature Human Behaviour*, 2:168–171, 2018.

[69] J. J. Lee and C. T. Chu. Bayesian clinical trials in action. *Statistics in Medicine*, 31:2955–2972, 2012.

[70] D. Lendrem, S. J. Senn, B. C. Lendrem, and J. D. Isaacs. R&D productivity rides again? *Pharmaceutical Statistics*, 14:1–3, 2015.

[71] T. Lewis. Professional development of statisticians in the pharmaceutical sector: evolution over the past decade and into the future. *Pharmaceutical Statistics*, 7:158–169, 2008.

[72] R. G. Marks, M. Conlon, and S. J. Ruberg. Paradigm shifts in clinical trials enabled by information technology. *Statistics in Medicine*, 20:2683–2696, 2001.

[73] D. V. Mehrotra, R. J. Hemmings, E. Russek-Cohen, and on behalf of the ICH E9/R1 Expert Working Group. Seeking harmony: estimands and sensitivity analyses for confirmatory clinical trials. *Clinical Trials*, 13(4):456–458, 2016.

[74] P. A. Milligan, M. J. Brown, B. Marchant, S. W. Martis, P. H. van der Graaf, N. Benson, G. Nucci, D. J. Nichols, R. A. Boyd, J. W. Mandema, S. Krishnaswami, S. Zwillich, D. Gruben, R. J. Anziano, T. C. Stock, and R. L. Lalonde. Model-based drug development: a rational approach to efficiently accelerate drug development. *Clinical Pharmacology and Therapeutics*, 93(6):502–514, 2013.

[75] J. Neyman. Inductive behavior as a basic concept of philosophy of science. *Review of the International Statistical Institute*, 25(1/3):7–22, 1957.

[76] A. O'Hagan and J. W. Stevens. Bayesian assessment of sample size for clinical trials of cost-effectiveness. *Medical Decision Making*, 21:219–230, 2001.

[77] J. O'Quigley, A. Iasonos, and B. Bornkamp. *Handbook of Methods for Designing, Monitoring, and Analyzing Dose-Finding Trials*. Chapman & Hall/CRC Press, Boca Raton, FL, 2017.

[78] J. O'Quigley, M. Pepe, and L. Fisher. Continual reassessment method: A practical design for phase I clinical studies in cancer. *Biometrics*, 46:33–48, 1990.

[79] J. Orloff, F. Douglas, J. Pinheiro, and et al. The future of drug development: advancing clinical trial design. *Nature Reviews Drug Discovery*, 8:949–957, 2009.

[80] P. Pallman, A. W. Bedding, B. Choodari-Oskooei, M. Dimairo, L. Flight, Hampson. L. V., J. Holmes, A. P. Mander, L. Odondi, M. R. Sydes, S. S. Villar, J. M. S. Wason, C. J. Weir, G. M. Wheeler, C. Yap, and T. Jaki. Adaptive designs in clinical trials: why use them, and how to run and report them. *BMC Medicine*, 16:29, 2018.

[81] C. R. Palmer and W. F. Rosenberger. Ethics and practice: Alternative designs for phase III randomized clinical trials. *Controlled Clinical Trials*, 20:172–186, 1999.

[82] V. Papadimitrakopoulou, J. J. Lee, I. I. Wistuba, A. S. Tsao, F. V. Fossella, N. Kalhor, S. Gupta, L. A. Byers, J. G. Izzo, S. N. Gettinger, and S. B. Goldberg. The BATTLE-2 Study: a biomarker-integrated targeted therapy study in previously treated patients with advanced nonsmall-cell lung cancer. *Journal of Clinical Oncology*, 34(30):3638, 2016.

[83] S. D. Patterson and B. Jones. *Bioequivalence and Statistics in Clinical Pharmacology, 2nd edition*. Chapman & Hall/CRC Press, Boca Raton, FL, 2016.

[84] A. N. Pitman and L. B. Pearce. A flexible outcome scoring system for clinical trials. *Clinical Pharmacology and Therapeutics*, 83(S1):S87, 2008.

[85] R. H. Randles and D. A. Wolfe. *Introduction to the Theory of Nonparametric Statistics*. Wiley, New York, 1979.

[86] L. A. Renfro and D. J. Sargent. Statistical controversies in clinical research: basket trials, umbrella trials, and other master protocols: a review and examples. *Annals of Oncology*, 28(1):34–43, 2017.

[87] C. Rosa, A. N. C. Campbell, G. M. Miele, M. Brunner, and E. L. Winstanley. Using e-technologies in clinical trials. *Contemporary Clinical Trials*, 45:41–54, 2015.

[88] W. F. Rosenberger and J. L. Lachin. *Randomization in Clinical Trials: Theory and Practice, 2nd edition*. Wiley, New York, 2015.

[89] W. F. Rosenberger, D. Uschner, and Y. Wang. The 15th Armitage lecture—Randomization: The forgotten component of the randomized clinical trial. *Statistics in Medicine*, to appear.

[90] S. L. Ruberg. Dose response studies I. Some design considerations. *Journal of Biopharmaceutical Statistics*, 5:1–14, 1995.

[91] SAS Institute. *SAS® 9.4 Language Reference: Concepts, 5th Edition.* Cary, NC: SAS Institute Inc., 2015.

[92] SAS Institute. *SAS® 9.4 Macro Language: Reference, 4th Edition.* Cary, NC: SAS Institute Inc., 2015.

[93] B. R. Saville and S. M. Berry. Efficiencies of platform clinical trials: A vision for the future. *Clinical Trials*, 13(3):358–366, 2016.

[94] J. W. Scannell, A. Blanckley, H. Boldon, and B. Warrington. Diagnosing the decline in pharmaceutical R&D efficiency. *Nature Review Drug Discovery*, 11:191–200, 2012.

[95] S. Senn. *Statistical Issues in Drug Development, 2nd edition.* Wiley, Chichester, West Sussex, England, 2007.

[96] R. Simon. Clinical trials for predictive medicine. *Statistics in Medicine*, 31:3031–3040, 2012.

[97] N. Ting, D. G. Chen, S. Ho, and J. C. Cappelleri. *Phase II Clinical Development of New Drugs.* Springer Nature Singapore Pte Ltd., 2017.

[98] H. Tobi, D. J. Kuik, P. D. Bezemer, and P. Ket. Towards a curriculum for the consultant biostatistician: identification of central disciplines. *Statistics in Medicine*, 20:3921–3929, 2001.

[99] R. L. Wasserstein and N. A. Lazar. The ASA's statement on p-values: Context, process, and purpose. *The American Statistician*, 70(2):129–133, 2016.

[100] R. L. Wasserstein, A. L. Schirm, and N. A. Lazar. Moving to a world "$p < 0.05$". *The American Statistician*, 73(S1):1–19, 2019.

[101] J. Whitehead and H. Brunier. Bayesian decision procedures for dose determining experiments. *Statistics in Medicine*, 14:885–893, 1995.

[102] N. Wiener. *God & Golem, Inc.: A Comment on Certain Points Where Cybernetics Impinges on Religion.* MIT Press, Cambridge MA, 1964.

[103] E. Wigner. The unreasonable effectiveness of mathematics in the natural sciences. *Communications in Pure and Applied Mathematics*, 13:1–14, 1960.

[104] Z. Williams, K. Roes, and N. Howitt. Qualified statisticians in the European pharma industry: Present and future directions. *Drug Information Journal*, 143:573–583, 2009.

[105] J. Woodcock and L. LaVange. Master protocols to study multiple therapies, multiple diseases, or both. *The New England Journal of Medicine*, 377:62–70, 2017.

[106] Y. Yuan, H. Q. Nguyen, and P. F. Thall. *Bayesian Designs for Phase I–II Clinical Trials*. Chapman & Hall/CRC Press, Boca Raton, FL, 2016.

[107] L. Zhang. *Nonclinical Statistics for Pharmaceutical and Biotechnology Industries*. Springer, Statistics for Biology and Health, 2016.

[108] X. Zhou, S. Liu, E. S. Kim, R. S. Herbst, and J. J. Lee. Bayesian adaptive design for targeted therapy development in lung cancer—a step towards personalized medicine. *Clinical Trials*, 5:181–193, 2008.

Index